AF707909

Modeling of Regional Atmospheric Pollution

SCIENCES

Climate and Atmosphere,
Field Director – Denis-Didier Rousseau

Atmospheric Physico-Chemistry, Subject Head – Benoit Laurent

Modeling of Regional Atmospheric Pollution

Laurent Menut

iSTE

WILEY

First published 2024 in Great Britain and the United States by ISTE Ltd and John Wiley & Sons, Inc.

Apart from any fair dealing for the purposes of research or private study, or criticism or review, as permitted under the Copyright, Designs and Patents Act 1988, this publication may only be reproduced, stored or transmitted, in any form or by any means, with the prior permission in writing of the publishers, or in the case of reprographic reproduction in accordance with the terms and licenses issued by the CLA. Enquiries concerning reproduction outside these terms should be sent to the publishers at the under mentioned address:

ISTE Ltd
27-37 St George's Road
London SW19 4EU
UK

www.iste.co.uk

John Wiley & Sons, Inc.
111 River Street
Hoboken, NJ 07030
USA

www.wiley.com

© ISTE Ltd 2024
The rights of Laurent Menut to be identified as the author of this work have been asserted by him in accordance with the Copyright, Designs and Patents Act 1988.

Any opinions, findings, and conclusions or recommendations expressed in this material are those of the author(s), contributor(s) or editor(s) and do not necessarily reflect the views of ISTE Group.

Library of Congress Control Number: 2023942929

British Library Cataloguing-in-Publication Data
A CIP record for this book is available from the British Library
ISBN 978-1-78945-102-3

ERC code:
PE10 Earth System Science
PE10_1 Atmospheric chemistry, atmospheric composition, air pollution

Contents

Preface

This text introduces the modeling of atmospheric pollution at a regional level. It goes back to the M1 (Masters 1) syllabus put together for the Magisterium at ENS Paris. As the physicochemical principles have already largely been introduced in other texts, the point of view for this book is the modeling of all these processes. Given that modeling is, in reality, a simplification within a particular framework, we will strive to demonstrate whether or not the processes to be represented, such as emission, chemistry and deposits, have been modeled well. There will be a discussion on the simplifying hypothesis used for each model, the ways in which these can be refined and the impact of these simplifications on the desired results. This text cites references that have been chosen for their relevance but also because these articles are free to access online through scientific journals. These references are therefore in *italics*. When they exist and are OpenAcess, the journal articles will be listed at the beginning of a chapter or a section. A list of all of these articles is given as follows:

– Atmospheric pollution, generalities and measurements:
  - *(Monks et al. 2015)*: https://acp.copernicus.org/articles/15/8889/2015/;
  - *(Benedetti et al. 2018)*: https://acp.copernicus.org/articles/18/10615/2018/.
– Gaseous chemical and aerosols:
  - *(Stockwell et al. 2012)*: https://www.mdpi.com/2073-4433/3/1/1;
  - *(Carslaw et al. 2010)*: https://www.atmos-chem-phys.net/10/1701/2010/;
  - *(Kanakidou et al. 2005)*: https://www.atmos-chem-phys.org/acp/5/1053/;
  - *(Fuzzi et al. 2015)*: https://acp.copernicus.org/articles/15/8217/2015/.
– Emissions:
  - Anthropic: *(Im et al. 2018b)*: https://acp.copernicus.org/articles/18/8929/2018/.
  - Biogenic:
    - *(Guenther et al. 2006)*: https://acp.copernicus.org/articles/6/3181/2006/;
    - *(Guenther et al. 2012)*: https://gmd.copernicus.org/articles/5/1471/2012/.
  - Biomass burning:
    - *(Reid et al. 2005a)*: https://acp.copernicus.org/ articles/5/799/2005/;
    - *(Paugam et al. 2016)*: http://www.atmos-chem-phys.net/16/907/2016/.

Modeling of Regional Atmospheric Pollution,
Laurent MENUT. © ISTE Ltd 2024.

- Mineral aerosols:
  - (*Schepanski 2018*): https://www.mdpi.com/2076-3263/8/5/151.
  - Volcanoes: https://acp.copernicus.org/articles/special_issue212.html.
  - Lightning: *(Schumann and Huntrieser 2007)*: https://www.atmos-chem-phys.net/7/3823/2007/.
  - Marine aerosols: *(Gantt and Meskhidze 2013)*: https://acp.copernicus.org/articles/13/3979/2013/.
  - Pollen: *(Sofiev et al. 2013a)*: https://www.ncbi.nlm.nih.gov/pmc/articles/PMC3527742/.

– Deposition:
  - Dry deposition:
    - *(Pleim and Ran 2011)*: https://www.mdpi.com/2073-4433/2/3/271;
    - *(Vivanco et al. 2018)*: https://www.atmos-chem-phys-discuss.net/acp-2018-104/.
  - Wet deposition: *(Zhang et al. 2019)*: https://doi.org/10.1029/2018JD029051.

– Modeling:
  - Current models: *(Baklanov et al. 2014)*: http://www.atmos-chem-phys.net/14/317/2014/.
  - Aerosol models: *(Semeniuk and Dastoor 2020)*: https://www.mdpi.com/2073-4433/11/2/156.
  - Data assimilation:
    - *(Sandu and Chai 2011)*: https://www.mdpi.com/2073-4433/2/3/426;
    - *(Bocquet et al. 2015)*: https://acp.copernicus.org/articles/15/5325/2015/.
  - Prediction: *(Marécal et al. 2015)*: https://gmd.copernicus.org/articles/8/2777/2015/.
  - Coupling:
    - *(Zhang 2008)*: https://acp.copernicus.org/articles/8/2895/2008/;
    - *(Baklanov et al. 2014)*: http://www.atmos-chem-phys.net/14/317/2014/.

There are other journal articles, but behind a paywall. These references are indicated in **bold**. They are cited directly within the relevant chapters. This text is also based on other books, including *(Jacob 1999)*, **(Delmas et al. 2005)**, **(Pielke 2013)**, **(Jacobson 1999)**, **(Stensrud 2011)**, **(Shao 2004)**, **(Seinfeld and Pandis 2006)** and **(Stull 1988)**.

For each chapter, the key points are summarized in gray boxes like this one.

November 2023

1

What is Atmospheric Pollution?

This chapter first presents how atmospheric pollution is defined. It then gives a succinct explanation of the principal pollutants, gases and aerosols, as well as the environment in which they will be modeled: the boundary layer and the troposphere. Our goal here is not to offer an exhaustive presentation of all processes related to atmospheric pollution, but to present an overview in order to better understand what can and cannot be modeled, and the limitations arising from the capacities of the models.

1.1. Introducing atmospheric pollution

1.1.1. *The concept of polluted atmosphere*

The concept of pollution, or a polluted environment, refers to any gaseous or particulate matter (the "pollutant") when its concentration exceeds the value that is commonly considered to be a mean value or a base value, or if this concentration presents a non-negligible impact on humans or the environment (**Delmas et al. 2005**). Consequently, the definition of pollution is very subjective, as it is related to the intensity, duration and potential impact. Pollution occurs across all levels of time and space: from pollution in urban regions, evolving from hour to hour, to long-term and global climate change. In the context of this text, the study of pollution is spatially limited to **a region**, or up to part of a continent: the **mesoscale**. The direct relations between space and time (atmospheric flux and lifetime of the chemical species) naturally result in a context where the processes in play are studied for **a few days to a few weeks**. In practice, studies have been carried out up to a few years (for studies on climate trends).

Modeling of Regional Atmospheric Pollution,
Laurent MENUT. © ISTE Ltd 2024.

Figure 1.1. *Industrial site in the 19th century (source: public domain: https://commons.wikimedia.org/wiki/File:Widnes_Smoke.jpg)*

Although the study of atmospheric pollution is one of the oldest problems (Figure 1.1), the awareness of the impact of human activity on the degradation of our own environment is more recent **(Fenger 2009)**. In general, as concerns studies on pollution, we can simply note here that the start of the 20th century marked an acceleration in our rate of destruction of the environment, the end of the 20th century saw a tepid awareness, and the start of the 21st century saw the first signs of attempts to slow down the pace of destruction. The major questions related to the study of atmospheric pollution are as follows: (i) *what impact do pollutants have on health?* (ii) *What impact does pollution have on the climate?* In both cases, there may be direct and rapid impacts (a peak in urban pollution or a volcanic eruption) and also slow and indirect impacts (illness linked to prolonged exposure to a pollutant, current global warming). In all of these cases, developing a chemistry-transport model could help us to generate answers as it is a **tool integrating all processes**. Similarly, only a numerical tool can predict these impacts. However, the model must have a certain level of maturity in order for the responses to be credible. Furthermore, this does not involve simply determining the concentration values of a pollutant at a particular time and place: we must also know the origins of this pollutant and the uncertainty attached to this value. From the early-1980s onwards, and following the development of powerful tools to model dynamic processes, work began on chemistry-transport models. Their development was based on the same logic as meteorological models: a complex tool for a complex system. The various models that existed at this time considered that the more processes there were, and the more finely represented they were, the better the results would be. This is not necessarily the case with chemistry-transport models, because the problem is different: their high sensitivity is not sensitive to the *initial conditions*. They are *models sensitive to boundary conditions*. Thus, it is not only important to start from the right point, but also to respect the balance between all of the forcings throughout the simulation.

1.1.2. *Where are the large polluted areas?*

A satellite provides a global image of the atmospheric concentration *(Georgoulias et al. 2019)*. This information is often integrated vertically, but shows the distribution of the largest emission zones and thus, in general, the most polluted areas. Figure 1.2 presents the measurements carried out by the GOME, SCIAMACHY and GOME-2 satellites for the period 1997–2017 and for NO_2. The regions with the highest emissions are clearly Europe, North America and China. If we zoom in on Europe, we can see that the regions with the highest NO_2 concentrations are the Benelux regions, the Po valley and most European supercities. We can also see the regions where biomass burning occurs, like the central regions of Africa. This figure also shows an analysis of the temporal variability of these NO_2 concentrations. Through the difference in measurements over this period, we can clearly identify the regions on the globe that have evolved, either in terms of a reduction or an increase in these concentrations: a trend that is a priori directly related to the intensity of surface emissions. The impact that China's development has had on the atmospheric composition can then be clearly seen. Global trends over the decades show a small decrease over North America and Europe, but a large increase over China.

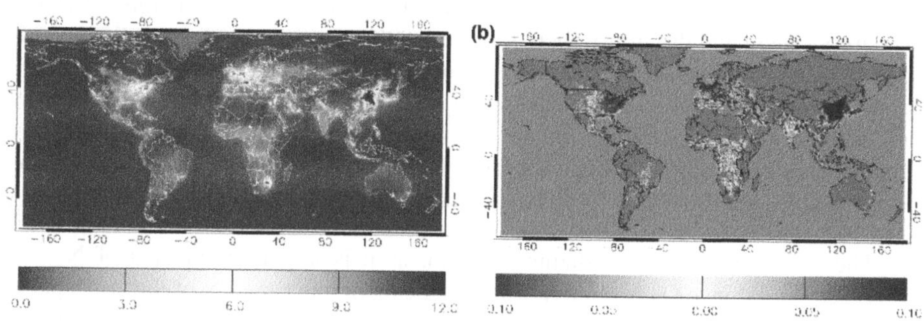

Figure 1.2. *(Left) Satellite measurements used to identify the global distribution of pollutants: integrated concentrations on the atmospheric NO_2 columns (10^{15} molecules cm^{-2}) for the period 1997–2017, measured by GOME, GOME-2 and SCIAMACHY. (Right) Trends for the integrated concentrations of NO_2 (10^{15} molecules cm^{-2} an^{-1}) in the same period. Figures adopted from* Georgoulias et al. (2019). *For a color version of this figure, see: http://www.iste.co.uk/menut/modeling.zip*

Historically, on a regional scale, the primary focus has been on the anthropogenic component: as the cause for the concentration, it therefore appears to be the most controllable component. Often linked to industries or vehicular traffic, this pollution is especially present in urbanized areas: air quality monitoring networks have thus started quantifying pollution in the spots where it was emitted and inhaled. One of the major pollutants that was "targeted" was **sulfur dioxide**. With this being greatly reduced today, attention has shifted to other gases such as **ozone and nitrogen oxides**. In more recent times, **particles** have become the focus. In parallel, although the phenomenon has always been present, there has been far less research into natural pollution. This is because it is not only noncontrollable, but has also been (wrongly) perceived as "background" pollution, which is less intense than anthropogenic pollution. In the 1980s, it was seen that pollution could travel. Just like other substances with a long lifetime (a topic of interest for climate studies), ozone and some of its precursors can be transported from one region to another. This means that local study, often in urban areas, is important, but it is also important to widen the zone of study: the air you breathe in a city could have been locally polluted, but could also have been polluted several hundred kilometers away, either by anthropogenic or biogenic causes. There is therefore a considerable increase in the uncertainty related to the composition and the ultimate fate of a mass of polluted air: just as rapid as the increase in the area being studied and the diversity of processes to take into consideration.

1.1.3. *Atmospheric composition*

The Earth's atmosphere is largely composed of gases and aerosols (**Jacob 1999**). The troposphere and the stratosphere represent 99% of atmospheric mass. Among the group of gases that constitute the atmosphere, the major natural greenhouse gases are water vapor and carbon dioxide. They play a particularly important role in the entire atmospheric system, including climate variability. At the tropospheric level, and on a time scale of a few days to a few weeks, the variability of the atmospheric composition is not affected very much by these two gases.

The atmosphere is an oxidizing medium, as it is largely composed of N_2 and dioxygen (O_2). Since O_2 is relatively stable, the oxidizing capacity of the atmosphere is especially dependent on photochemical reactions. These reactions are initiated by solar irradiation, which leads to the production of radical species. Atmospheric chemistry is therefore a process that causes primary species to tend toward a higher state of oxidation, that is, acidification. The atmospheric composition is driven by a biogeochemical cycle made up of **sources** (the emission of gas and particles), **transformations** (gas and particulate chemistry, transport, mixtures) and **sinks** (dry and wet depositions). These transformations are summarized in Figure 1.3.

The reactions exist in three forms: (i) oxidation transforms gaseous compounds into other gaseous compounds. A "kinetic process" is when gaseous reactants collide with one another and create a new gaseous product. Photolysis is the process where a gaseous reactant is "broken up" by solar irradiation. Photochemistry encompasses all kinetic processes in gaseous phase and photolysis. (ii) This can then be followed by the conversion of gas to particles through various processes, such as nucleation, which is the aggregation of gas molecules into a small particle, and then condensation when a gas aggregates around a nucleus and changes to liquid state, and evaporation when a liquid molecule again becomes gaseous at the surface of a particle. (iii) Finally, heterogeneous and aqueous phase reactions will directly cause the gas to interact with a drop of water or a natural aerosol that has already been formed (mineral dust, sea salt).

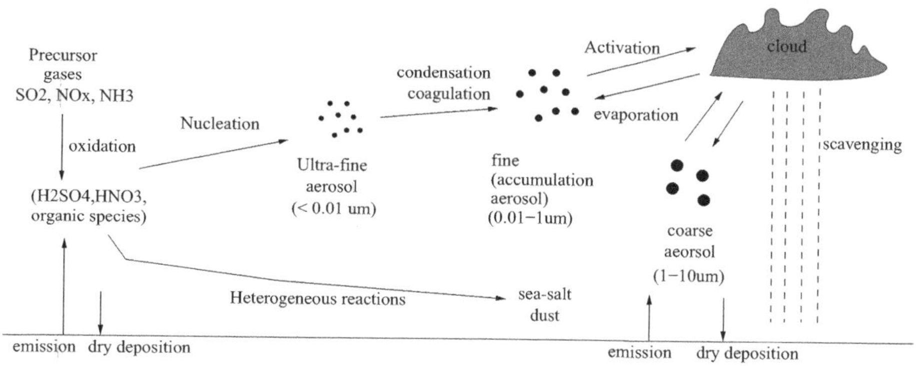

Figure 1.3. *Lifecycle of gases and aerosols. Figure adapted from* **Jacob (1999)**

Many chemical species are present in the troposphere and those of particular interest with regard to atmospheric pollution are presented in Figure 1.4. These species are classified into families based on their role and lifetime. The species with the shortest lifetimes (a few seconds or minutes) are **oxidizing agents**. They are present in atomic and molecular form (ozone, peroxyacetyl nitrate (PANs), etc.) or in the form of radicals (OH, HO_2, RO_2, NO_3, etc.). Their chemical action occurs principally through **photochemical reactions**. The species with a lifetime of the order of an hour to a year are **reactive gases**. These are carbon derivatives (carbon monoxide and volatile organic compounds (VOCs)), nitrogen derivatives (nitrogen oxide, ammoniac, etc.) and sulfur derivatives (sulfur dioxide, hydrogen sulfide, etc.). These gases are monitored with respect to atmospheric pollution. Among the species whose lifetime is greater than a year are **greenhouse gases**. These absorb infrared radiation and influence the Earth's radiative balance. These gases are, mainly, water vapor (H_2O), carbon dioxide (CO_2), methane (CH_4), nitrogen protoxide (N_2O), ozone (O_3) and chlorofluorocarbons (CFCs). They have been studied especially in the climatological context.

There are different units used to quantify the abundance of gases and aerosols in the atmosphere. The units can differ based on the source of the data, the measurements or the modeling used. First of all, gases can be quantified using the **mixing ratio**.

The mixing ratio C_x is the number of moles of the gas divided by the number of moles of air (1 mole = N_A = 6.022 $\times$ 10^{23} atoms, Avogadro's number). This is a dimensionless quantity and is invariant, regardless of the density of air. We then use *ppb* (part per billion) or *ppm* (part per million) to express this very small ratio. It is also possible to quantify gases using **number density**. This unit, n_x, represents the number of molecules per unit volume of air. This makes it possible to relate gas density with the quantities required for optical calculations, notably by vertically integrating over the atmospheric column. This quantity is not conserved when pressure or temperature vary. This unit is in *molecules per* cm^{-3}. Finally, concentration can be quantified by **mass**, m_x. This represents the mass of the pollutant per unit volume and is expressed in μg.m^{-3}. This quantity is related to n_x through the molar mass M_x of the gas (kg.mol^{-1}) and Avogadro's number N_A, such that $m_x = n_x M_x / N_A$.

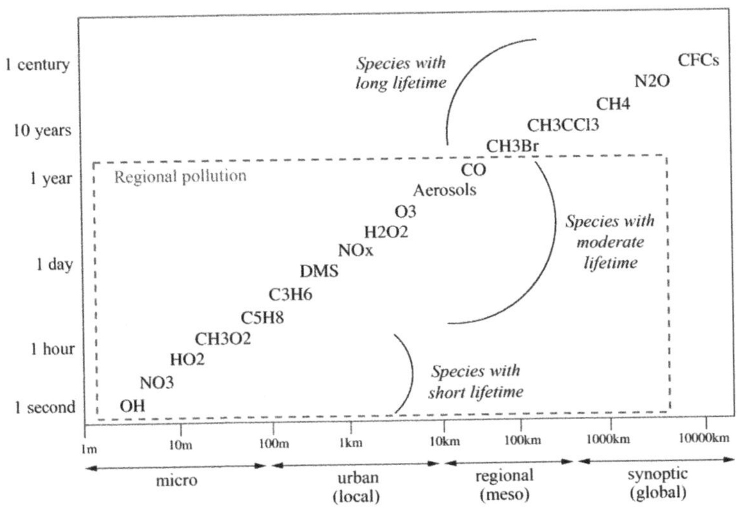

Figure 1.4. *Chemical species in the atmosphere and their lifetimes. Adapted from* **Seinfeld and Pandis (2006)**. *For a color version of this figure, see: http://www.iste.co.uk/menut/modeling.zip*

All of these "pollutants" will reside for a greater or shorter duration in the atmosphere, based on their physical or chemical nature, as well as their position on the planet (altitude and latitude) and therefore on the meteorological conditions in their environment. To evaluate and quantify their persistence in the atmosphere, it is possible to evaluate their residence time and their characteristic transport times. If we consider that this residence time is most dependent on global atmospheric circulation, it is possible to arrive at an initial idea of these characteristic transport times. The general circulation of the atmosphere is driven by mechanisms such as the vertical thermal instability of the atmosphere, the rotation of the earth and the inhomogeneities in the distribution of solar energy. The characteristic transport time is presented in Figure 1.5. We can see that the time for iso-latitude transport is a few days to a few months, while iso-longitude transport time is around a year. Therefore, for species with a long lifetimes (such as CO_2 and CH_4), we arrive at a homogenization of concentrations in the upper atmosphere (except the boundary layer) as their lifetime is much longer than their transport time.

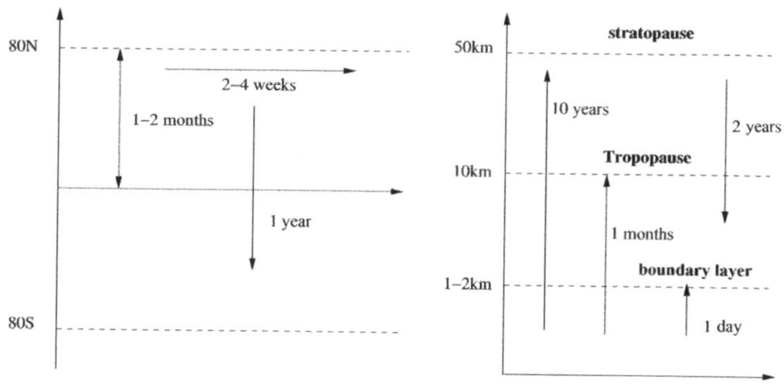

Figure 1.5. *Characteristic transport time. Figure adapted from* **Jacob (1999)**

1.2. Typical concentrations by season

We will present the principal characteristics of episodes of winter or summer pollution (knowing that autumnal and spring episodes have both characteristics) for a continental region, with highly urbanized areas. For each of these periods, the peaks in pollution will be related to the meteorological conditions and the different pollutants **(Seinfeld and Pandis 2006)**.

Episodes of summer pollution are first and foremost photochemical episodes, leading to the production of ozone downstream of urbanized centers. Ozone peaks are created when conditions of low wind, high temperatures and strong sunlight persist over several days. This is the case with "anticyclonic periods". During these periods, we have both high anthropogenic emissions of NO_x and VOCs, which stagnate close to the emission site, as well as biogenic VOC emissions and active photolysis. Winter episodes are more related to nitrogen oxides and particles. Anthropogenic emissions are greater in winter, especially from traffic and residential heating. This is combined with two effects: meteorology shows lower winds and less active turbulence. There are lower temperatures and less sunlight, limiting the production of secondary pollutants such as ozone. The background ozone (apart from the peaks) will vary between 30 and 100 μg m^{-3} through the year. NO_x concentrations are half of this and vary between 10 and 50 μg m^{-3} through the year, with a maximum in winter.

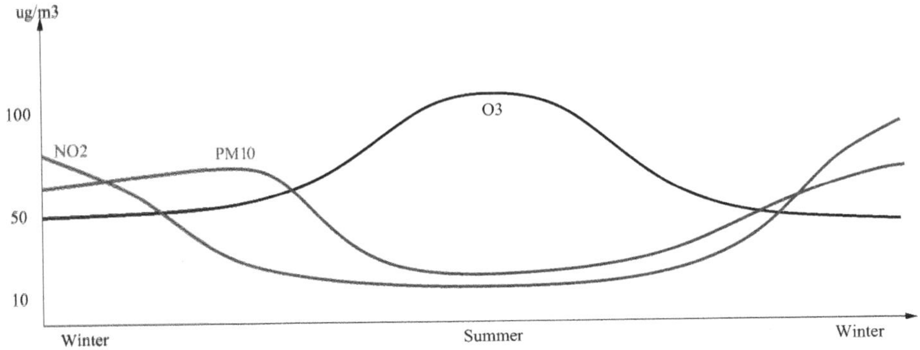

Figure 1.6. *Seasonality of pollutants. Ozone is produced through photochemistry and has a seasonal peak in the summer. NO_x and PM_{10} are produced more so in the winter. A secondary PM_{10} peak may appear in the spring, with agricultural fertilizers. For a color version of this figure, see: http://www.iste.co.uk/menut/modeling.zip*

1.3. Gas-phase chemistry

Open-access publications: *Stockwell et al. (2012).*

1.3.1. *Abundance and lifetime of species*

Gas-phase chemistry involves hundreds of chemical species and reactions **(Seinfeld and Pandis 2006)**. The major chemical species concerned are listed in Table 1.1. To be more precise, the species involved in the pollution in the boundary layer are highlighted in gray. It can be seen that these species have a lifetime of a few days to a few weeks. In terms of concentration, the relative abundance is low, leading to the name "trace gases" **(Jacob 1999)**.

The principal atmospheric oxidizing agents are present in atomic and molecular form (ozone, PANs, etc.) or as radicals (OH, HO_2, RO_2, NO_3, etc.). These are species with a very short lifetime and their action occurs primarily through photochemical reactions. The oxidation process leads to carbon being transformed into CO_2, hydrogen into H_2O, nitrogen into HNO_3 and sulfur into H_2SO_4. The action of the oxidizing agent O_2 (molecular oxygen) has a preponderant role to play in the final balance. Atmospheric gas-phase species are classified into several distinct families based on their molecular structure.

Inorganic gases contain any element, single or associated with a molecule of hydrogen (H) or carbon (C). For example, ozone (O_3), the hydroxyl radical (OH), water vapor (H_2O), nitrogen dioxide (NO_2), sulfur dioxide (SO_2) and carbon monoxide (CO). **Organic gases** contain any element plus hydrogen (H) and carbon (C) molecules. For example, we have alkanes (methane (CH_4), ethane (C_2H_6)), alkenes (ethene (C_2H_4), propene (C_3H_6)), aromatics (alcohols, terpenes, aldehydes, including formaldehyde (HCHO)) and nitrates (methyl nitrate (CH_3ONO_2)). **Hydrocarbons** are organic gases that only contain H and C molecules, for example, alkanes and alkenes. Non-methane hydrocarbons are all hydrocarbons, except methane, which is considered to have very low reactivity at the level of atmospheric pollution. Gaseous species can also be differentiated based on origin. **Primary pollutants** are emitted directly into the atmosphere. For example, we have the species NO and NO_2 (NO_x), SO_2 and the VOCs. **Secondary pollutants** are formed from primary pollutants and during chemical reactions: ozone (O_3), a gas that is not visible but is a respiratory irritant; nitrogen dioxide (NO_2), which is notably produced by automobiles, is a colored pollutant with a pungent odor; formaldehyde (HCHO) is produced by smoke from forest fires, exhaust fumes from automobiles and tobacco, which is also called *formalin*; nitric acid (HNO_3) is found in acid rain and is also called *aqua fortis*; finally, sulfuric acid (H_2SO_4), which is also found in acid rain and is called *vitriol*.

Component	Formula	% volume/dry air	Residence time
Water vapor	H_2O	Variable	10 days
Dinitrogen	N_2	78.084	15 million years
Dioxygen	O_2	20.948	8000 years
Carbon dioxide	CO_2	0.037	15 years
Methane	CH_4	1.7×10^{-4}	9 years
Hydrogen	H_2	5×10^{-5}	10 years
Carbon monoxide	CO	1.2×10^{-5}	2 months
Ozone	O_3	2 to 200×10^{-6}	1 to 2 months
Ammoniac	NH_3	0.1 to 10^{-6}	20 days
Nitrogen dioxide	NO_2	10^{-7}	1 day
VOCs	$C_xH_yO_z$	0.1 to 10^{-6}	A few hours
Sulfur dioxide	SO_2	2.10^{-8}	1 day
Hydrogen sulfide	H_2S	2.10^{-8}	1 day

Table 1.1. *Gas-phase chemical species in the atmosphere*

Gas-phase species are also classified based on their action. **Greenhouse gases** absorb infrared radiation and affect the Earth's radiation balance. This is the case with water vapor (H_2O), carbon dioxide (CO_2), methane (CH_4), nitrogen protoxide (N_2O), ozone (O_3) and chlorofluorocarbons (CFCs). **Reactive gases** are carbon derivatives (carbon monoxide and VOCs), nitrogen derivatives (nitrogen oxides, ammoniac, etc.) and sulfur derivatives (sulfur dioxide, hydrogen sulfide, etc.).

1.3.2. *Ozone and the radical cycle*

Ozone is a chemical species that plays a variety of roles, with different impacts. Whether in the stratosphere or the troposphere, ozone is produced by the recombination of an oxygen atom and O_2. It is one of the principal pollutants that is monitored with respect to air quality. While the destruction of the ozone layer is a problem in the stratosphere (studies carried out on the "hole in the ozone layer"), at a regional level and in the troposphere, it is the high production of ozone that poses a problem, especially close to the surface in the boundary layer. Ozone is produced in the troposphere through the oxidation of CO and VOCs by OH and in the presence of NO_x (= NO + NO_2). This entire process is called the **radical cycle**. This cycle primarily occurs during the day: photolysis produces free radicals in an atmosphere that is highly chemically reactive. On the contrary, at night, without sunlight, there are very few radicals and therefore there is no active chemistry. The principal oxidizing agents that play a role in this cycle are the hydroxyl radial (OH) (produced by photolysis during the day), ozone (O_3) (produced by photolysis during the day) and the nitrate radical (NO_3) (consumed by photolysis during the day).

1.3.3. *Chemical regimes*

In highly anthropized regions, the question that arises is knowing which emissions must be reduced for the maximum effect on ozone concentrations **(Honoré and Vautard 2000)**. Since the principal precursors are NO_x and VOCs, it is useful to find indicators for understanding which of these two sources must be reduced on priority. As the chemistry vis à vis these two products is nonlinear, a reduction in one or the other will not necessarily lead to a reduction in ozone **(Sillman 1999)**. *Chemical regimes* are defined by considering the relative part of the NO_x and VOC emissions upstream of the region being studied. Figure 1.7 presents the iso-values of ozone (curves, called *isopleth diagrams*) based on NO_x and VOCs emissions. In the upper region to the left of the diagonal, called the *limited VOC regime*, a reduction in NO_x emissions (with constant, low VOC emissions) leads to an increase in ozone. This regime is typical of an urbanized environment. On the contrary, an increase in VOCs will increase ozone levels. This is the case in northern Europe **(Beekmann and Vautard 2010)**. In the bottom right section of the diagram, called *limited* NO_x *regime*, ozone increases when NO_x increases, but is relatively unaffected by changes in VOC levels. This is more typically the case in rural areas **(Seinfeld and Pandis 2006)**. Modeling these chemical regimes does not require any particular development in a model: the production dynamic of ozone, like other species, based on emissions, is implicitly taken into account in the chemical mechanisms used. The only question that arises today is the influence of the resolution of the model. This is because, based on the size of the mesh, highly localized NO_x sources can be more or less diluted, thus changing the VOC/NO_x ratio.

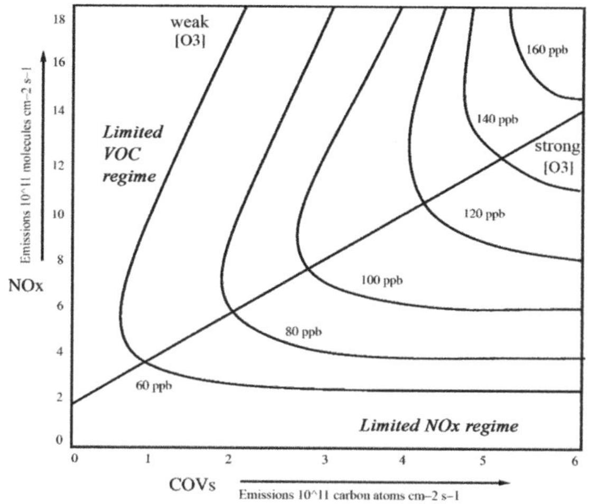

Figure 1.7. *Isopleths characterizing chemical regimes: the evolution of ozone concentrations based on the relative emissions of NO_x and VOCs. The values given are indicative in nature and represent orders of magnitude*

1.3.4. *Typology of urban pollution*

A "typical" episode of pollution in an urban setting is schematically depicted in Figure 1.8. The concentration values are indicative and are only meant to show the diurnal evolution of each type of chemical species. We can differentiate between concentrations with a single peak (ozone) and with double peaks (NO_x and VOCs) in a day. The double peaks are related to urban emissions: NO_x and VOCs are primary species. Since NO_x emissions are largely from traffic, the time of these peaks corresponds to the peaks in urban traffic. VOCs are also emitted by traffic and by urban heating: we therefore have a cycle with nonzero values throughout the day. In the morning, the time difference between NO and NO_2 can be explained by the primary emissions of NO, which are then rapidly converted to NO_2. VOCs are oxidized to oxide radicals and aldehydes (R-CHO, with R being a chain containing carbon, for example: ethanal (CH_3CHO)). At the end of the forenoon, the decrease seen in concentrations of NO_x and VOCs is due to a triple effect related to their consumption, the end of their emission and the development of the convective boundary layer (and therefore, the vertical dilution of these species). In the afternoon, NO_2 is converted to NO through photolysis. The molecular oxygen recombines with O_2 to form ozone. However, the concentrations of this ozone are lower in the city than in the outskirts or countryside, as the chemical regime promotes its titration by NO. Aldehydes and NO_2 form PAN. This PAN will accumulate during the night and can be transported over long distances: it is a distant source and reservoir of NO_x.

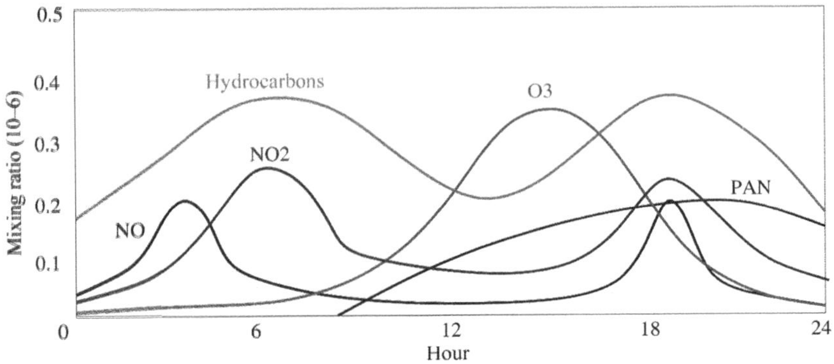

Figure 1.8. *A "typical" episode of pollution in an urban setting. For a color version of this figure, see: http://www.iste.co.uk/menut/modeling.zip*

1.4. Aerosol chemistry

An aerosol is a solid or liquid particle in suspension in the atmosphere and with a *mean mass median diameter* (MMD) ranging from ≈ 2 nm to ≈ 100 μm. It is generally made up of a set of particles with a solid kernel and a liquid or gaseous envelope. For the same aerosol species, we can have particles of different sizes. When the molecule groups are made up of gaseous pollutants, we can precisely quantify their size, mass and shape. This is not the case for aerosols that may be made up of different species, with very different sizes and shapes. Although they are not necessarily spherical, aerosols can be classified based on their MMD. These aerosols are also denoted by *particulate matter* (PM). The group of aerosols with MMD smaller than 2.5 and 10 μm are thus categorized as $PM_{2.5}$ and PM_{10}, respectively **(Seinfeld and Pandis 2006)**.

Aerosols may be directly emitted into the atmosphere (primary aerosols) or formed through gas-to-particle conversion or the condensation of gaseous components around preexisting fine particles (secondary aerosols) **(Monks et al. 2009)**. **Primary aerosols** are largely composed of **soots** formed by the combustion of hydrocarbons, **minerals** produced by the stripping away of the Earth's surface and **marine salts** produced by sprays from the surface of the sea. **Secondary aerosols** are divided into two different families: (i) inorganics, with sulfates, nitrates and ammonium NH_4^+, and (ii) organics, when they are made up of C, O, H, N and S, among which we find hydrocarbons (C_xH_y), alcohols (R-OH), ethers (R-O-R'), aldehydes (r-CHO), acetones (R-CO-R') and carboxylic acids (R-COOH) *(Kanakidou et al. 2005)*. As shown in Figure 1.3, the aerosol formation cycle can be broken down into several processes: (1) nucleation, (2) condensation and (3) coagulation. Many parameters come into play and are most often uncertain. The first process is the oxidation of gaseous components into semi-volatile components. Based on the volatility, we then have a transfer between the gaseous-phase and particle phase. This transfer occurs, first through nucleation, corresponding to the creation of kernels through the transfer from the gaseous state to the liquid/solid state in a homogeneous manner. Then we have condensation, which is the transfer from the gaseous state to the particle state. This transfer can take place either through adsorption, that is, chemistry between the aerosol surface and the gas-phase molecule, or through absorption, that is, the dissolving of the gas-phase molecule onto the (liquid) surface. In terms of speciation (i.e. the relative abundance of the different chemical species), it is seen that we find an equivalent mass of inorganics and organics. This is only true if we consider all aerosols, without discriminating based on size. If we differentiate between fine aerosols and coarser aerosols (the boundary being a median diameter of 2 μm), the speciation will be different. In the fine mode, the aerosol is primarily made up of sulfates, ammonium and organic matter. On the contrary, in the coarser mode, there is a predominance of natural primary aerosols: mineral dust, sea spray and metals.

1.4.1. *Abundance of aerosols*

The map in Figure 1.9 shows the average concentration of aerosols on the surface of the globe. These values correspond to $PM_{2.5}$ concentrations. Their origin is not specified and the "mix" that is presented has aerosols related to pollution (anthropogenic), biogenic aerosols or natural ones. It must be noted that the major part of the aerosols are found in the Northern hemisphere. The highest mass values, corresponding to mineral dust aerosols, are produced by erodible areas (deserts in Africa, Saudi Arabia, India and China). Smaller but significant concentrations can be seen in industrialized regions: North America, Europe and China. Finally, Central Africa also has non-negligible concentrations due to biomass burning. These aerosols are all accounted for in pollution measurements, but they have very different characteristics. In order to evaluate their lifecycle, we will need to quantify their emissions, transport and deposition.

1.4.2. *Size distribution*

Aerosols result in very different size distribution from the emission stage onwards and these also evolve over the course of their lifecycle. However, despite the wide diversity, several "modes" are systematically repeated. Figure 1.10 presents two types of size distributions used for measurements and modeling: distribution based on number and based on mass. For these distributions, we see peaks in the abundance of aerosols, and these correspond to the processes that are to be modeled. Nucleation is the creation of aerosol kernels for gases which generate very small particles of the order of $D_p < 10$ nm. The Aitken mode represents the condensation and coagulation processes. These are aerosols of size $\mu m < D_p < 0.1\ \mu m$. The accumulation mode represents aerosols of size $0.1\ \mu m < D_p < 1\ \mu m$. This is also a coagulation and condensation mode, which, in this case, can lead to the formation of droplets. Finally, the "coarse" mode corresponds chiefly to primary aerosols such as minerals.

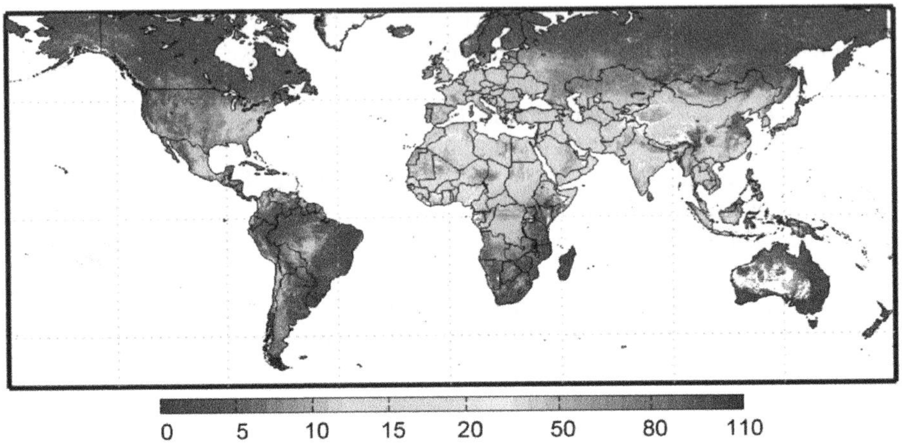

Figure 1.9. *Aerosols with diameter smaller than 2.5 μm across the globe (Snider et al. 2015). For a color version of this figure, see: http://www.iste.co.uk/menut/modeling.zip*

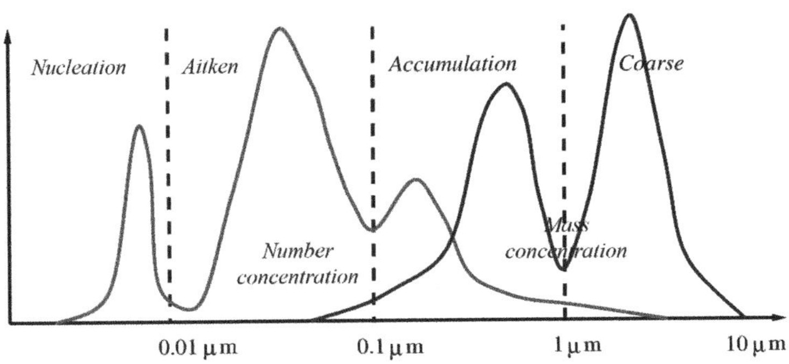

Figure 1.10. *Typical shape and size distribution of aerosols. For a color version of this figure, see: http://www.iste.co.uk/menut/modeling.zip*

1.5. The atmospheric boundary layer

The atmospheric boundary layer (ABL) is the seat of most of the sources and wells of the atmospheric components that we study. It is also the site where we live and, therefore, of the air we breathe. It is a volume of air that changes based on time of day as well as location. When modeling pollution, the knowledge of this volume will govern the mixture of pollutants emitted on the surface: the accuracy of its value will thus be very sensitive, as an error of 10% in this volume will directly produce an error of 10% in the concentrations. Both the dilution volume as well as the intensity of the turbulent mixture in this volume must be modeled in order to provide the best vertical profile of pollutants in the atmospheric basin.

1.5.1. *General structure*

The ABL consists of the interface between the Earth's surface and the free troposphere. It is the lower part of the atmosphere, which is directly influenced by terrestrial processes **(Stull 1988)**. It extends vertically from the Earth's surface to a few hundreds or thousands of meters of altitude and depends directly on all meteorological parameters (wind, temperature, humidity, sun radiation) as well as the topography and type of environment (oceanic, rural or urban continental). The ABL is thin (1–2 km) in comparison to the rest of the troposphere ($\approx$ 11 km) or compared to the whole of the atmosphere. The concept of the thickness of the ABL has no instantaneous physical reality: unlike other variables that are much more tangible, such as wind or temperature, the ABL thickness is a statistical result. A temperature can be instantaneously measured, but we can only take the average of physical quantities to deduce a mean value for the height of the ABL.

The principal processes taking place within the boundary layer are the transport of momentum, heat and humidity. These processes are, above all, fluxes, turbulent processes. The processes within the ABL are therefore always described and studied using a mean-term and a turbulent term.

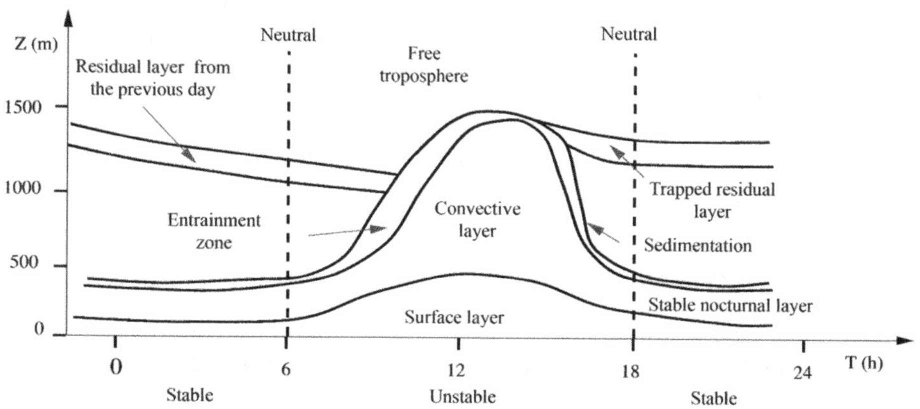

Figure 1.11. *Schema showing the temporal evolution of the boundary layer. The times and altitudes given are indicative. For a color version of this figure, see: http://www.iste.co.uk/menut/modeling.zip*

The ABL evolves with a diurnal periodicity, whose phases are invariant in principle, but whose amplitude depends on dynamic and thermal processes related to the site being studied. Different parts of the ABL are defined depending on the hour, day and altitude. The **surface layer** is present throughout the diurnal cycle and its thickness is approximately 10% of the mean height of the ABL, denoted by $\overline{h}$. In the case of **mixed layer**, under the influence of the heating of the surface, the boundary layer is developed mainly as a function of vertical stability, from when sunlight begins. The ABL is therefore the site of intense turbulent activity, having thermal and dynamic origins. The **drag zone** is relatively thinner (from some tens of meters to a hundred meters) and makes up the summit of the boundary layer and is the interface between the ABL and the free troposphere. It must be noted here that we use "free troposphere" because unlike the ABL, the upper part of the troposphere is not subject to the influence of terrestrial processes. The **residual layer** appears at the end of the day. In the absence of dynamic forcings, the aerosols within the mixed layer will either sediment toward the surface or remain trapped in the residual layer. This residual layer may be stable or remain turbulent (especially in an urban setting). Finally, there is the **stable nocturnal layer**, which is established after sunset. However, in an urban setting this may still be neutral or even unstable in some cases (due to the effect of anthropogenic nocturnal heating, highly local turbulent processes such as air-recirculation between buildings).

As with any fluid medium, the ABL is particularly sensitive to phenomena occurring at its boundaries: the base (the surface layer) and its summit (defined by the drag zone, in the convective period, marking the start of the free troposphere). Thus, for similar surface fluxes, large-scale conditions will influence the evolution of the ABL. Over a high-pressure zone (such as the center of an anticyclone), we have the subsidence of masses of air. On the surface, therefore, there is the divergence of wind. At the summit of the ABL, we often observe cumulus clouds. Anticyclonic conditions are conducive to the creation of situations with high atmospheric pollution. In a low-pressure zone, the boundary layer is much lower and winds converge. Above, convective systems are established. These are linked to strong updrafts and are conducive to the emergence of stormy conditions.

1.5.2. *Diurnal cycle and turbulence*

The variability of the boundary layer on the meso-scale must therefore be studied temporally, both horizontally and vertically **(Stull 1988)**. In addition to the topography, type of terrain (plains, sea, forest etc.), the most intense exchanges take place along the vertical in the boundary layer. Therefore, for a given site, and all throughout the day, it is possible to characterize the boundary layer by studying the vertical profiles. Figure 1.12 presents a few profiles from different hours in the day. A particularly useful variable, in this case, is the potential temperature, θ :

$$\theta(z) = T(z) \times \left(\frac{P_0}{P(z)}\right)^{R/C_p} \qquad [1.1]$$

with P_0 being the reference surface (by definition, equal to 1,000 hPa), g being the acceleration due to gravity, R being the perfect gas constant, such that R = 287.05 J kg^{-1} K^{-1} and C_p= 1,005 J kg^{-1} K^{-1} being the calorific capacity of the air. This potential temperature corresponds to the temperature that a parcel of air would have if it was brought adiabatically (without any heat exchange with its environment) from its pressure $P(z)$ to the pressure P_0. In addition, we also define the virtual potential temperature θ_v, as the potential temperature taking into account the effects of humidity, such that:

$$\theta_v(z) = \theta(z)\,(1 + 0.61q) \qquad [1.2]$$

with q being the specific humidity (g kg^{-1}).

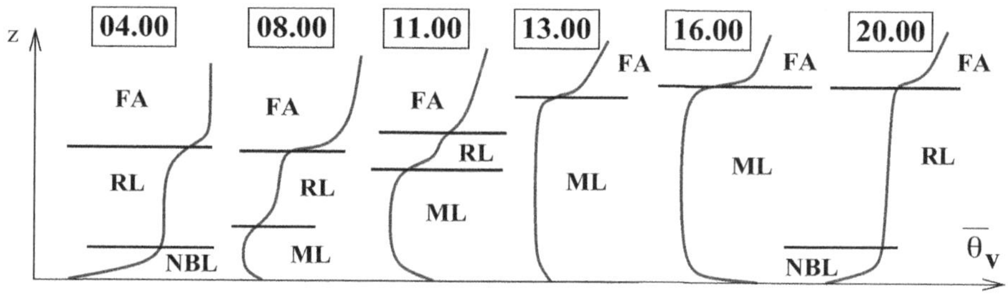

Figure 1.12. *Diurnal evolution of the profiles of the virtual potential temperature $\overline{\theta_v}$ within the atmospheric boundary layer (ABL). The notations used are as follows: NBL: nocturnal boundary layer; SBL: surface boundary layer; RL: residual layer; FA: free atmosphere; ML: mixed layer. For a color version of this figure, see: http://www.iste.co.uk/menut/modeling.zip*

For each of these profiles, we can identify the vertical structure of the ABL at different times of the day (the timezone being UTC). At *04.00*, θ increases in the surface layer, also called the stable nocturnal layer (T decreases). The layer is stable. In the residual layer, θ is constant, and we therefore have a neutral layer. Then, above the free atmosphere, we observe a positive vertical gradient, typical of the atmosphere on a larger scale, and almost constant. In general, this gradient, denoted by γ is of the order of 6.5 K/km and, under our latitudes, it does not do very much. On the contrary, for a given altitude, a variation in the value of γ is representative of the advection of a mass of hotter or cooler air on the synoptic scale. At *08.00*, the only difference from the preceding hour is the appearance of the mixed layer on the surface, which replaces the stable layer. The sun begins to heat the ground, and we obtain a negative gradient of θ: the layer is unstable and any parcel can rise in altitude through the phenomenon of buoyancy. At *11.00*, the mixed layer continues to extend vertically and is aggregated into the residual layer, as in the case of atmospheric pollution for example. It is at this time that the residual layers of ozone, which until now had been trapped at altitude, are reabsorbed into the mixed layer. At *13.00*, we are in what is called the convective period, that is, vertical processes largely prevail over horizontal processes. The atmosphere is unstable or neutral: any air parcel on the surface will rise as far as the free troposphere. In these cases, we observe structures said to be "thermal structures", which are rapid lifts. Typically, a parcel of air can move from the surface to an altitude of 2,000–3,000 m in less than 15 minutes. At *16.00*, the difference widens on the surface and the gradient grows steeper and steeper. At this time of maximum sunlight, the ABL achieves its highest altitude. At *20.00*, the sun no longer heats the surface and ground temperatures fall very quickly (slower in the city). We quickly shift from a negative gradient on the surface to a positive gradient: the heaviest pollutants (such as dust) will sediment in the surface layer, while the lightest pollutants (such as ozone) may remain trapped in the residual layer.

Depending on the hour of day, the boundary layer can be turbulent, leading to quicker or slower mixing of the pollutants emitted and formed close to the surface. The intensity of the mixing may completely change the concentration of the pollutants measured on the surface. This turbulence is characterized based on the stability or instability of the boundary layer. These instabilities are characterized by the relative part of the thermal and mechanical thermal production. We thus define the flux Richardson number, which is the ratio between the flow shear term and the buoyancy term, such that:

$$R_{if} = \frac{\frac{g}{T}\overline{w'\theta'}}{\overline{u'_i u'_j}\frac{\partial \overline{u_i}}{\partial x_j}} \qquad [1.3]$$

with $\overline{w'\theta'}$ and $\overline{u'_i u'_j}$ being the turbulent heat fluxes and momentum, respectively; u_i being the wind component in the direction i, and the indices being i, j or k for the directions x, y or z. With $R_{if} > 0$, we have a stable atmosphere; with $R_{if} = 0$, a neutral atmosphere, and with $R_{if} < 0$, we have an unstable atmosphere. We also commonly use the *bulk Richardson number* (R_{ib}), which is calculated based on the vertical flux gradients.

1.5.3. *The urban boundary layer*

As the development of the boundary layer is first and foremost driven by surface characteristics, the urban boundary layer (UBL) is highly specific **(Barlow 2014)**. As the urban environment is the site of the major part of anthropogenic emissions, it is primordial to study it. The modeling of the UBL is a research topic that has seen great development in the past few decades. In this introductory section, we will only present an outline of the major characteristics that we wish to model. The modeling itself will be presented in greater detail in Chapter 3.

1.5.3.1. *Characteristics*

The characteristics of the UBL were described in the 1980s, notably in the landmark publications by **McRae et al. (1982)** and **Oke (1982)**. The concept of the "urban heat island" was introduced in these, with the earliest energy balance calculations. At this time, the modeling could not accurately account for this environment and the parameterizations were only slightly modified with the appropriate increases in roughness and heat flux **(Pielke 2013)**.

Figure 1.13 depicts a schematic view of a city surrounded by a suburban environment (the suburbs) in order to understand the phenomenology of what must be modeled. The differences in the ABL between the urban and rural environments are primarily due to the surface characteristics of both of these environments: the composition of the ground (grass, tar) and therefore its capacity for storing heat and humidity; and the composition of the surface (forests, fields, buildings) with different roughnesses influencing atmospheric flux. The albedo and emissivity (the ground's capacity to store and emit thermal radiation) are lower in the city than in the urban environment, notably due to the presence of tarred surfaces. Furthermore, the large presence of aerosols in an urban environment tends to reduce incident solar flux by an order of 10–20% (depending on the site of study). Cities generally have a higher temperature than the countryside due to heating, small industries, etc. This phenomenon is accentuated in winter, which leads to the definition of an urban heat island **(Oke 1982)**. The net impact is an additional heat flux, called an anthropogenic heat flux, which is added to the radiative balance in the environment: in winter, we find values that can go up to 30 W/m^2. The wind speed will be lower in the urban environment than the rural environment in the day, due to the presence of high buildings that slow it down. On the contrary, it is higher at night: the effects of the high horizontal temperature gradients become dominant. An analysis of urban air shows that it is generally drier than in a rural environment, which is again due to the type of surface and the coating on the surface. This difference is not limited to the atmosphere: the urban underground is "hollow", due to pipes and sewers, which is rarely the case in open lands.

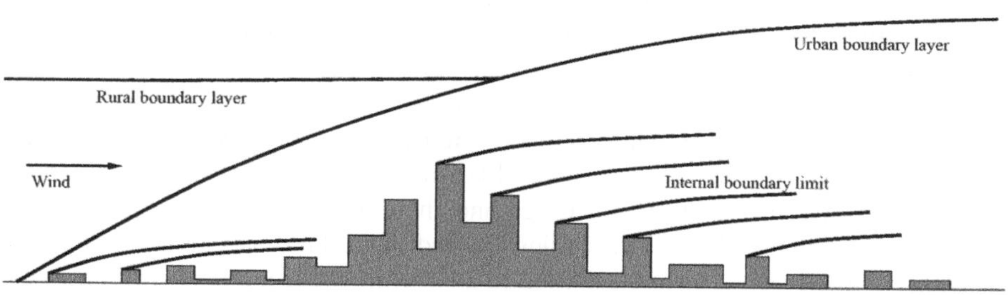

Figure 1.13. *Structure of the urban boundary layer*

We therefore see differences in how humidity may be stored underground (for a shorter duration, in cities). In certain conditions, the urban environment may reproduce phenomena studied in nonurban environments, such as the development of an internal boundary layer and land breeze. On a more regional scale, many synoptic phenomena may significantly influence the development of the urban convective boundary layer. These influences can be summarized as follows: (i) an advection of cold air above the convective ABL will accelerate its development, an advection of hot air will slow it down; (ii) steep pressure gradients on the surface will create strong winds; (iii) low stability will accelerate the mixing processes; and (iv) weak winds (above the ABL) will reduce the transfer through vertical movements. In an anticyclonic period and with weak winds, an advection of hot air above a city will thus slow down the growth of the ABL, and trap all aerosols and pollutants, leading to an increase in urban pollution.

1.5.3.2. *Urban modeling*

It was not until the 1990s that we could represent these phenomena, which were both regional and local: simulations improved in resolution and models became capable of explaining circulation around an urban center, as well as the disturbances induced, such as the increase in the precipitation regime downstream of the urbanized zone **(Baik and Chun 1997)**. In addition to taking into account the disruption in the atmospheric flux at a regional level, caused by the city, accounting for turbulence and vertical mixing was also adapted to better represent the urban environment. Since large differences were observed at night, the parameterizations were modified **(Kot and Song 1998)**. For larger-scale models, adapting the turbulence in a mesh was no longer enough: the urban environment has an impact, but its surface represents only a small percentage of the surface. It was thus necessary to implement subgrid representations, as in **Calbo et al. (1998)**, who developed statistical adaptations for urban emission of pollutants based on high-resolution simulations. In the 2000s, schemes that integrated all urban processes to respect the energy balance were developed and integrated into meso-scale models: they statistically described the effects of streets and buildings, as in the *Town Energy Budget* (TEB) scheme in **Masson (2000)**, used in the Meso-NH model to analyze the urban thermal breeze in the Parisian region in **Lemonsu and Masson (2002)**. In the same year, the scheme by **Martilli et al. (2002)** was proposed, which would be integrated into the WRF model (see p. 43). After purely meteorological studies, accounting for urban construction made it possible to quantify its impact on atmospheric pollution as well **(Sarrat et al. 2006; Auger and Legras 2007)**. In the 2010s, statistical schemes like the TEB were enhanced by adding more details on heat flux from buildings and their impact on flow, as mentioned in **Krpo et al. (2010)**. Sensitivity studies showed that taking into account submesh variability in the urban environmental led to an overall change in results **(Flagg and Taylor 2011)**. In parallel, meteorological studies on a local scale (i.e. one hundred meters) were carried out using computational fluid dynamics (CFD) tools in order to explicitly quantify the impact of buildings on atmospheric flux and turbulent fluxes

(Milliez and Carissimo 2008; Fesquet et al. 2009; Burman et al. 2019). All of these developments in modeling, often based on adapting parameterization for a larger scale, were because of dedicated measurement campaigns, such as the ESCOMPTE project in the Marseille region **(Mestayer et al. 2005)**, FUMAPEX in Copenhagen **(Baklanov et al. 2008)**, Salfex in Salford (UK) **(Barlow et al. 2008)** and Joint Urban in Oklahoma City **(Finn et al. 2010)**. At present, studies on urban environments (whether they are experimental, using CFD, high-resolution studies, through submesh variability, or statistical parameterization studies) are still being actively carried out *(Buccolieri and Hang 2019)*. The challenge that remains is to model meteorology and concentrations of pollutants, and also the exposure of populations for studies on the health impacts of atmospheric pollution.

2

Observations and Legislation

Observations are an essential part of analyzing any physicochemical process. They represent the only reality, with modeling being only an interpretation of this reality. In this chapter, we will describe the different types of meteorological and chemical measurements that are of interest to understand an episode of atmospheric pollution. We will discuss the feasibility of the measurements, their accuracy and their representativity. Their use in a model, in the form of data validation or assimilation, is discussed in Chapter 7 (p. 149). Whether in meteorology or chemistry, there are several different types of measurements. First of all, we have *in situ measurements*: these are very locally representative, but only at the particular site and time of measurement. However, they have the advantage of being more realistic as they correspond directly to the information that we are looking for. We then have *remote detection measurements*: these make it possible to access information at a distance. There are two subgroups within these remote data: measurements in the form of profiles (the entire distance that is scanned is filled using spatial sampling) or in an integrated form (there is only one mean value across the measurement path).

Depending on the variable being measured and the instrument used, there will be a large variability in the *uncertainty* and the *representativity*: two concepts that must be quantified if we want to compare measurements with the modeling results, or use them for data assimilation. For the same type of instrument and variable, the interpretation of the measure will depend on the temporal frequency of acquisition, the scope and the spatial resolution of the instrument and of the type of variable measured.

If you want to follow research into meteorology and atmospheric composition, there are many different journals, listed in Table 2.1. Finally, in the context of the simulation of pollution, and with a chemistry-transport model, we will be led to compare the results using both meteorological and chemical measurements.

Modeling of Regional Atmospheric Pollution,
Laurent MENUT. © ISTE Ltd 2024.

Journal	Open Access
Journal of Atmospheric and Oceanic technology www.ametsoc.org/ams/index.cfm/publications/journals	very limited
Remote Sensing www.mdpi.com/journal/remotesensing	yes
Atmosphere www.mdpi.com/journal/atmosphere	yes
Atmospheric Measurements Techniques www.atmospheric-measurement-techniques.net	yes

Table 2.1. *Examples of scientific journals on atmospheric observation*

2.1. Meteorological measurements

Meteorological measurements make it possible to understand the nature of a particular situation and to then understand whether conditions for an episode of pollution are present. This also makes it possible to quantify the relevance of the main forcing in a modeling of atmospheric pollution: dynamic and turbulence.

2.1.1. *In situ measurements*

In situ measurements are the only measurements that represent reality as they are directly acquired from the site where we are looking for information. However, they have highly limited spatial and temporal representativity and therefore generally cannot be compared with a modeling result. These measurements can be divided into two categories: (i) routine network measurement and (ii) occasional measurements (measurement campaigns). **Routine network measurements** offer the advantage of several locations with the same measure, temporal frequency and representativity. Instruments are intended for automatic use and data acquisition. They must therefore be robust and the measure must not derive (too) much over time. Among these routine measurements, we have (i) surface stations, (ii) measurement towers and (iii) radiosonde. **Surface stations** are automatic stations that transmit data through a dedicated network. The acquired data require very little post-processing; however, it is often reserved for the organizations that deploy the instruments and are not made public. The stations measure many variables through different instruments: temperature (at 2 m), humidity (at 2 m), pressure (at 2 m), wind (at 10 m), radiation (integrated), nebulosity (integrated), etc. The data obtained is hourly. These regular and continuous-time measurements make it possible to establish long time series and "reframe" a particular period with respect to a wider period (a given week contrasted with the whole season, for example). It must however be noted that the wind data are only recorded with a precision of the order of meters per second: as situations of pollution often occur in low-wind situations, the variability in the measurements is low.

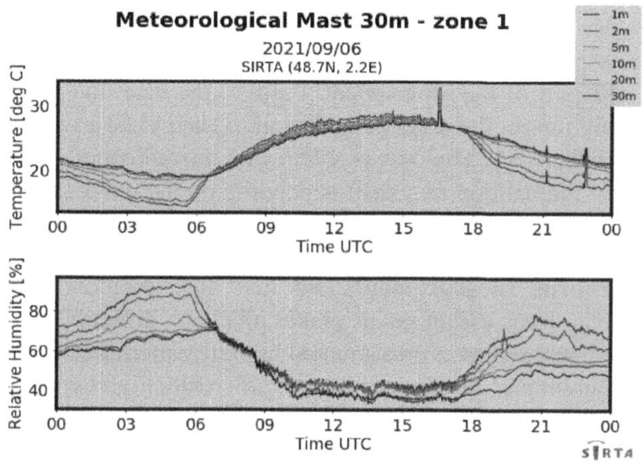

Figure 2.1. *Temperature and relative humidity at different heights on the 30 m mast on the instrumented SIRTA site in Palaiseau. For a color version of this figure, see: http://www.iste.co.uk/menut/modeling.zip*

Measurement masts (or towers) are also considered to be routine measurements. Unlike surface stations, the measurements are carried out at varying heights above ground level (even if the "norm" is to measure between 10 and 30 m above the ground). These measurements bring in important additional information: the concept of the vertical gradient. If the data are acquired at a high temporal frequency, it makes it possible to access turbulence measurements. Thus, we find **sonic anemometers** on the masts. It works on the principle of measuring wind fluctuations at high frequency through cups in three directions. We thus directly obtain the values for u', v' and w'. This makes it possible to directly derive the friction velocity u_* and the turbulent kinetic energy (TKE). In general, these measurements serve as the standard for turbulence measurements and make it possible to "align" measurements from other instruments. An example of the measurements is presented in Figure 2.1. It shows the measurement of temperature and relative humidity on the 30 m high instrumented mast, at the *Site Instrumental de Recherche par Télédétection Active* (SITRA, Instrumental Site for Research Through Active Remote Detection) in Palaiseau. The variability of these quantities can be seen based on a small variation in height, with respect to the ground. The SIRTA group encompasses a set of measurements from the lower atmosphere – these are both meteorological and chemical (https://sirta.ipsl.fr/).

Radiosondes produce both routine measurements (depending on networks) or measurement campaigns. They are balloons, launched several times a day, from different stations and as they ascend vertically, they provide measurements for wind, temperature, pressure and humidity. On average, a balloon takes between 30 and 40 minutes to rise from the surface to an altitude of 20 km. The data from these radiosondes are especially valuable because they provide the only real, routine in situ measurement of a part of the atmosphere. Further, the data that are recorded have a high vertical resolution (of the order of a few meters) and yield a variety of information on stratification in the ABL. The only problem with the radiosonde is that its trajectory is not controlled – the balloon may drift depending on the wind and altitude. Thus, in order to interpret these data, we must take the trajectory into account.

Dynamic **airborne measurements** are first and foremost used to carry out measurements over long distances in places that are difficult to access (sea and mountains). Wind, temperature, pressure and humidity measurements are carried out. When these measurements are carried out at a sufficiently high temporal frequency, we can determine their mean value and standard deviation. Related to the uncertainty of the instrument, it is then possible to deduce turbulent fluxes from this. In practice, these measurements serve to constrain and validate a model, or to perfect new parametrizations.

2.1.2. *Measurements through remote detection*

Remote detection measurements make it possible to have wider spatial and temporal cover. However, they require measurements to be interpreted to obtain geophysical data. Indeed, the signal that is received represents an effect, but it cannot precisely quantify the cause: these measurements make it possible to improve knowledge about a physical phenomenon, but they must be accompanied by uncertainties with respect to the interpretation of the measurement, in addition to the uncertainty linked to the representativity of the measurement. We present here instruments used to analyze the atmosphere close to the surface: **lidar**, **radar** and **sodar**. These measurements are carried out at the **zenith** or from the ground, in any direction with an angle incident, that is, from the surface to the atmosphere. For the lidar, there are also airborne measurements at both the zenith and the **nadir**. Beyond the atmosphere, we will also present **satellite** measurements.

Lidar (*light detection and ranging*) is composed of a laser source (laser – *light amplification by stimulated emission of radiation*), which emits a beam into the atmosphere. As it propagates, this beam will encounter particles and thus be attenuated and also reflected (or backscattered). A part of this backscattered signal will return toward the ground. A telescope is placed close to the laser and receives this signal *(Piironen and Eloranta 1995)*. Analyzing this signal using the Doppler effect in its emission frequency band will provide a vertical backscatter profile. After correcting this signal, and based on the emission wavelength, it is possible to access the vertical profile of the presence of clouds or aerosols. One of the advantages of this measure is its high acquisition frequency (generally about 10 Hz) and vertical resolution (a few meters of resolution over the whole troposphere). The drawbacks are that the signal is often noisy, the measure must be inverted to arrive at geophysical quantities and the profile is limited to the site where the instrument is located. The diffusion of the laser emission is said to be elastic if the backscattered frequency is the same as the emitted frequency (such as Mie or Rayleigh diffusion), and it is said to be inelastic if it changes. Rayleigh diffusion is therefore the scattering by atoms and molecules, while Mie diffusion is the scattering by aerosols or cloud droplets. There are also Raman lidars, with scattering by molecules and inelastic emission, with the change in frequency being a signature of the molecule that was encountered. The principal types of lidar used are as follows: backscatter, DIAL (*differential absorption lidar*), Raman and Doppler. The backscatter lidar measures the backscattered radiation and polarization. DIAL measures the concentration of chemical species by emitting in two wavelengths. It is possible to estimate the concentration by comparing the absorbent wavelength and the non-absorbent wavelength for the molecule being studied. Raman lidar can also measure the targeted chemical species by measuring the wavelength modification in the received signal. Lidars are fragile and costly equipment, requiring an operator and large-scale maintenance. In recent years, companies have innovated and offer smaller lidars that can be embedded in aircraft or cars. Depending on the type of laser, the type of measurement and inversion algorithms, we can obtain, using the lidar, vertical concentration profiles of aerosols, ozone or wind *(Baumgarten 2010)*. One of the most common applications is determining the height of the ABL (**Menut et al. 1999; Hennemuth and Lammert 2006; Morille et al. 2007**). In Europe, lidars are often implemented in networks, with the *European Aerosol Research Lidar Network* (EARLINET, see: https://www.earlinet.org) *(Pappalardo et al. 2014)*. This makes it possible to obtain homogeneous measurements at different sites in Europe and allows us to track a transport event over a long distance, for example.

Sodar (*sound detection and ranging*) is also based on the backscattering principle, not based on a laser, like lidar, but an audio sequence with different wavelengths (**Singal 1993**). We then obtain backscattering profiles, and also (as there are three distinct antennae) the wind measurement u, v and w. Sound is propagated less than light in the atmosphere, with the obtained measurements only covering a distance of 500 or 600 m from the ground. The advantages of this instrument are, above all, its ability to measure **wind profiles**, as well as the possibility of reconstructing **turbulent parameters** along the vertical. In practice, this instrument has proven to be tricky to use when there is a lot of "background noise" (as in the urban milieu). However, it is still regularly used for studies of the urban boundary layer, especially in winter when this layer remains low and therefore within the scope of the instrument (**Dupont et al. 1999; Pichelli et al. 2014**). The baseline measurement of this instrument is the backscattered signal S, which is related to the turbulent fluctuations of temperature and wind:

$$S^2\, exp(2az) = b_1 \frac{C_T^2}{T^2} + b_2\, C_V^2 \qquad [2.1]$$

with T being the temperature, a being the attenuation coefficient and z being the distance between the sodar antenna and the target that has been reached. The quantity C_T^2 (n K^2 $m^{-2/3}$) characterizes the intensity of the small-scale inhomogeneities in the temperature field, and b_1 and b_2 are proportionality constants. When the transmitter and the receiver are confounded, which is generally the case, only the temperature fluctuations contribute to the backscattering (i.e. the C_V^2 term is zero). Sodar has proven to be particularly useful in the study of turbulence in the surface layer. Using the standard deviation measurements for vertical velocity σ_w, parameterizations allow us to obtain time series of sensible surface heat flux *(Weill et al. 1979)*.

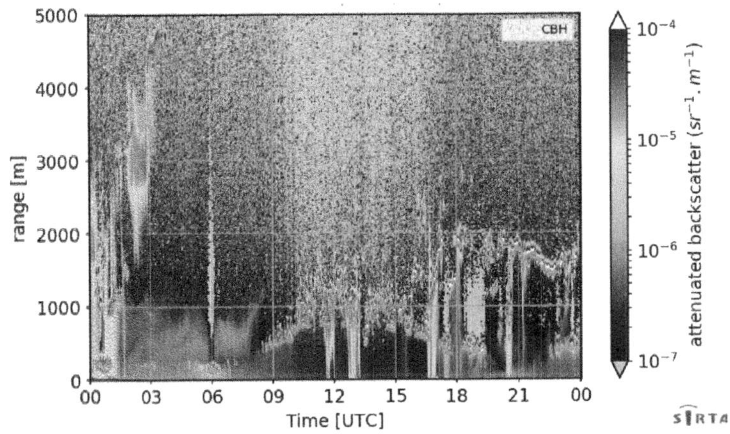

Figure 2.2. *Vertical cross-section of a corrected lidar signal on the SIRTA site. Figure based on the public database: https://sirta.ipsl.fr/data_quicklooks.html. For a color version of this figure, see: http://www.iste.co.uk/menut/modeling.zip*

Radar functions on the emission of an impulse at a regular interval. The antenna makes it possible to measure the part of the signal that is backscattered on water droplets, in the case of precipitation. It allows us to characterize precipitation in three dimensions, with high spatial or temporal resolution. Figure 2.3 presents a vertical cross-section of the vertical velocity taken from a radar echo during a day in October 2020. The boundary layer is low and materialized by the presence of clouds at a constant altitude of 1,500 m in the day.

From the ground, different instruments allow us to obtain measurements in the incident radiation flux. **Photometers** measure the light radiation in wavelengths such that it is perceptible to human vision. Several wavelengths can be measured and a joint analysis makes it possible to obtain information on the nature of the clouds or aerosols present in the atmosphere. **Pyranometers** (solar radiation, *short-wave radiation*) are thermal flux sensors. They measure the power of the total solar radiation in W.m^{-2}. A **pyrgeometer** is a hemispherical radiometer. It measures the incident infrared radiation (*long-wave radiation*) due to Earth radiation originating in the atmosphere or the ground. These instruments are not cumbersome and are automatic. They allow continuous measurements to be carried out without an operator.

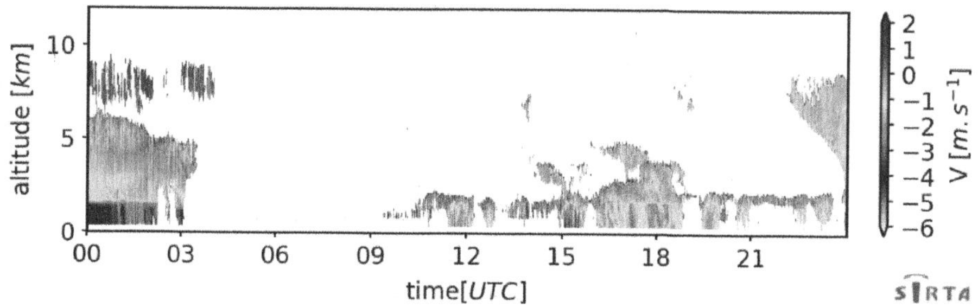

Figure 2.3. *Vertical cross-section of the radar signal on the SIRTA site. Fall velocity in hydrometeors. Figure based on the public database: https://sirta.ipsl.fr/data_quicklooks.html. For a color version of this figure, see: http://www.iste.co.uk/menut/modeling.zip*

2.2. Atmospheric concentration measurements

Open-access publications: *Benedetti et al. (2018).*

As with meteorology, measurements of chemical concentrations in the atmosphere are organized into a network or one-off measurements. They can also be in situ measurements or remote detection measurements **(Laj et al. 2009)**.

2.2.1. *In situ measurements*

In situ measurements are mainly carried out by automatic surface stations connected in a network, with routine operations. As with meteorological measurements, we have access to concentrations on an hourly frequency and for fixed locations. The networks are largely installed and maintained by accredited air-quality monitoring agencies. These agencies directly use their measurements for local analyses, however they also share them in real-time at national and international levels. In France, these measurements are managed by the *Laboratoire Central de*

Surveillance de la Qualité de l'Air (LCSQA, Central Laboratory for Air Quality Monitoring). At the European level, these measurements are then entered into public databases with the European organization European Monitoring and Evaluation Programme (EMEP) (Figure 2.4) and the Airbase run by the European Environment Agency (EEA). There are more measurements at the local level than at the European level, as EMEP stations are a sub-sampling of measurements corresponding to rural measurements. Unlike in meteorology, the tracking of the measured concentrations is carried out in a legislative framework: regions are obliged to monitor air quality through the measurement of concentrations. In order to do this, the network must carry out measurements uninterrupted, and the representativity of the measurements must be known. Therefore, AASQA has a large network of stations, as well as other measurements carried out in parallel to complement this network. (i) Measurements in mobile vehicles allow us to determine the variability and therefore the representativity of a measurement around a station. (ii) Measurements in their meteorology department allow these researchers to calibrate the measurement instruments in order to reduce the bias due to deviations.

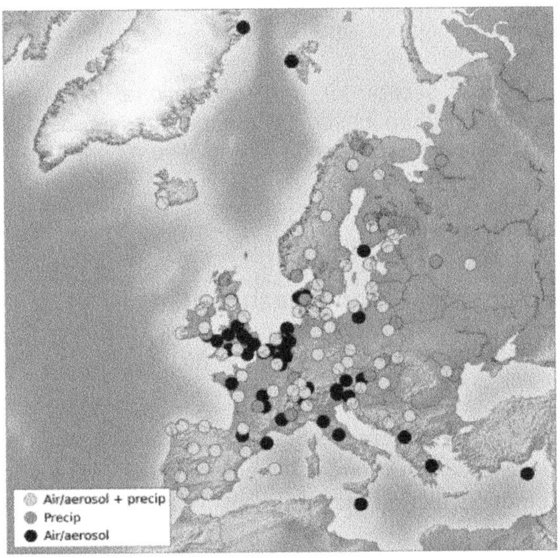

Figure 2.4. *Stations in the measurement network implemented and managed by the European EMEP program. For a color version of this figure, see: http://www.iste.co.uk/menut/modeling.zip*

Different instruments have been developed and are used depending on the pollutant that is to be measured. Among the techniques used by networks or measurement campaigns, we have, first, measurement through UV fluorescence or emission spectroscopy. The measurements are continuous and the detection limit is of the order of a few ppb. This instrument is used, for example, for measuring SO_2. The principle it operates on is UV radiation of gas molecules, passing to an excited state and then relaxed toward their fundamental level. This generates a photon emission from which the concentration is deduced. In order to measure NO, NO_2 and O_3, instruments based on the chemiluminescence principle are used. The principle is also the emission of light energy, the result of the return to a fundamental state by the product of a chemical reaction. For this to be done, the desired gas is brought into contact with an excess of a compound, so that the chemical reaction is complete. We can also look at the differential optic absorption spectroscopy (DOAS): radiation upon a molecule produces spatially heterogeneous scattering, called *Rayleigh scattering (Gratsea et al. 2020)*. This scattering depends on the effective section of absorption of the gas and on the incident wavelength. Recent studies use many instruments of this type in a network in order to make comparisons with similar measurements obtained through satellites *(Verhoelst et al. 2020)*.

Figure 2.5 presents an example of a measurement carried out at SIRTA in Palaiseau with an *Aerosol Chemical Speciation Monitor* (ACSM). Developed by a private company, this is not a prototype, but remains an instrument reserved for the specific purposes of measurement and research. This is an adaptation of instruments based on the principle of the *aerosol mass spectrometer* (AMS). It allows us to quantify concentrations of ammonium sulfate, ammonium nitrate, organic matter and carbon soot in the range of very fine sizes, with PM_1 and $PM_{2.5}$ (therefore, with a mean median mass smaller than 1 and 2.5 μm, respectively).

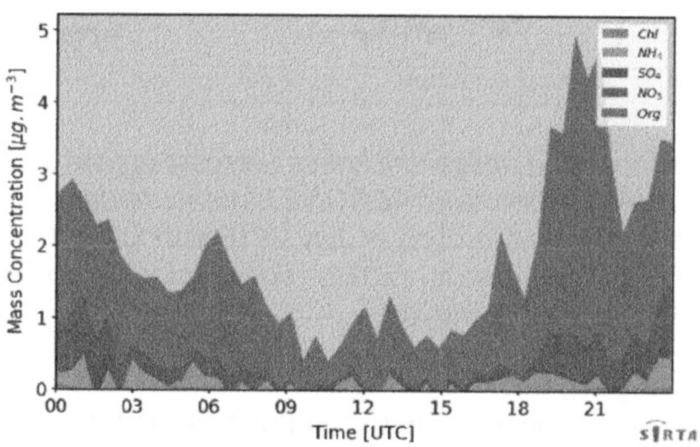

Figure 2.5. *ACSM measurements at SIRTA. Aerosol composition during the day of October 3, 2020, in the Paris region. It can be seen that the organic fraction of the aerosol is the dominant part. For a color version of this figure, see: http://www.iste.co.uk/menut/modeling.zip*

2.2.2. *Measurements through remote detection*

There are two types of remote measurements: (i) surface measurements that vertically probe the atmosphere and (ii) satellite measurements.

2.2.2.1. *Measurements from the surface*

For measurements from the surface, the main global network of interest is the Aerosol Robotic Network (AERONET)[1] **(Holben et al. 1998; Dubovik and et al 2000; Sinyuk et al. 2020)**. It enables us to obtain values related to aerosols: optical thickness (integrated over the entire atmospheric column) for different wavelengths and which is only due to aerosols (an algorithm eliminates cloud contribution) and the computation of the Angström coefficient. The measurements are not carried out at regular times, but are almost hourly. The network is spatially very dense and studying the time series or maps allows us to see if a model has accurately returned the transport of a dense plume of aerosols through comparison between stations **(Kinne et al. 2003)**. The data from this network are widely used for modeling studies, for example in the Mediterranean basin, to quantify the veracity of their results, as seen in some previous studies *(Bergamo et al. 2008; Basart et al. 2009; Menut et al. 2016).*

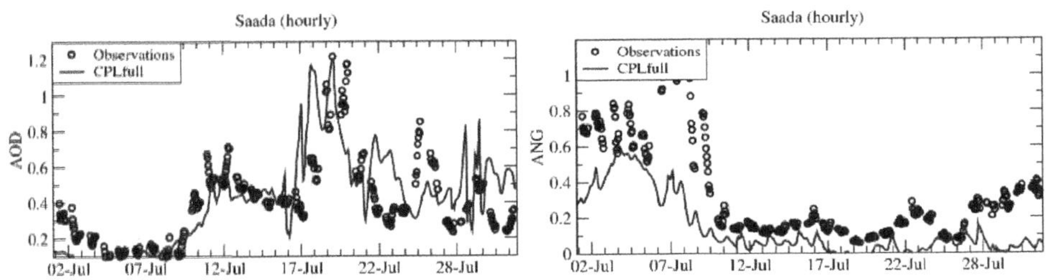

Figure 2.6. *Comparison of the optical thickness (aerosol optical depth (AOD)) and the Angstrom coefficient (ANG) between AERONET measurements and the CHIMERA model for the Saada station in Morocco in July 2016 during the DACCIWA campaign. The data are taken from Menut et al. (2019). For a color version of this figure, see: http://www.iste.co.uk/menut/modeling.zip*

Figure 2.6 presents a comparison between the AERONET measurements and the CHIMERE model for the Saada station in Morocco, during the month of July 2016. At the beginning of this period, the AOD is very low, showing either few aerosols or aerosols that are not very optically active. The Angström shows observed values of 0.8, which the model under-estimates (0.6). This is a fine aerosol and therefore of anthropogenic origin. An increase in the optical thickness toward the middle of the month, up to a value of 1.2, is a sign of the transport of a large aerosol mass.

1 AERONET: http://aeronet.gsfc.nasa.gov.

The corresponding Angström shows small values, smaller than 0.2, therefore a coarse aerosol: these are mineral aerosols, in this case. Globally, the inter-day variations are well reproduced, including the phasing of peaks, showing that the model is able to reproduce emissions and their transport.

2.2.2.2. *Satellite measurements*

In recent years, satellite measurements have increased in number and quality. Satellite measurements for the chemical composition of the troposphere were first used on a global scale. As they have low spatial and temporal resolution, and are more qualitative, they have made it possible to describe the transport of plumes of pollutants over long distances for species that have an atmospheric lifetime of a few days. The improvement in measurement and progress made in algorithms for analysis today have made it possible to have precise information on a regional, or even metropolitan scale. By covering zones that have very few, or no in situ measurements, satellite observations can be used at various levels: they can help constrain pollutant emission inventories, make it possible to have a fine analysis of simulation results over vast regions and over any atmospheric column, or again, be assimilated into models to improve their performance.

There are many sensors, each of which has technical characteristics that make it possible to discriminate one or many atmospheric chemical species, with a variety of measurement frequencies and spatial coverage (Table 2.2). From this information, we get a broad picture of the advantages and drawbacks of the existing sensors used to study regional atmospheric pollution. The majority of the pollutants we wish to quantify are today measured by at least one sensor. However, while the quantification of gaseous species has reached a high level of precision, this is not the case for aerosols. These are often only represented in the form of their effect on the optical thickness (AOD), without discriminating between the species that are present, nor looking at the particle size distribution. On the contrary, the variety of satellites allows us to refine the assessment of a concentration through the overlap between measurements.

Among the satellites that can be used for atmospheric composition, we differentiate between LEO and GEO satellites. LEO (*Low Earth Orbiters*) are satellites that travel at a low altitude. They offer the advantage of a finer vertical resolution than the GEO (geostationary) satellites, with global coverage (but with legs that vary in width), and of sensors that can be used for chemical species that are of interest for pollution. On the other hand, they only pass over a given site once a day and they have low sensitivity in differentiating between the ABL and the rest of the troposphere. GEO, or geostationary, satellites are in orbit at a fixed point of 36,000 km from Earth. Their fixed position allows acquisitions at regular intervals of time and, therefore, the possibility of creating time series, as well as maps for a given region. On the contrary, they only take observations in one region and have no vertical resolution. The horizontal resolution remains low due to their longer distance. Finally, these satellites are primarily dedicated to meteorology, and at present there have been no measurements directly linked to atmospheric pollution, in the sense of relatively accurate measurements of concentrations close to the surface. Among the orbiting satellites, we have the *A-Train* satellite constellation coordinated by NASA. It is composed of different satellites, each dedicated to a specific geophysical measure: Cloudsat for cloud reflectivity; CALIPSO for aerosols **(Winker et al. 2009)**, clouds and ice; AIRS/AQUA for temperature profiles, the pressure at the top of the clouds, CO concentration *(Kopacz et al. 2010)*; TRMM for precipitation **(Cecil et al. 2014)**; AURA for ozone, water vapor, CFCs, methane, nitrogen; and PARASOL, which uses the POLDER radiometer (*POLarization and Directionality of the Earth's Reflectance*) for surface characteristics **(Deschamps et al. 1994)**. The IASI sensor on the Metop platform, initially designed for meteorological measurements (temperature), also makes it possible to estimate chemical concentrations in the troposphere: O_3, CO, CO_2, H_2O, CH_4. Close to 1.3 million profiles are measured every day. To arrive at vertical concentration profiles, we must "invert" the data by using profiles called *averaging kernels (Clerbaux et al. 2009)*.

The advantage of satellite observations is that measurements can be carried out across the globe, regardless of the installation of a measurement network on the ground. One of the drawbacks is that we often only have measurements integrated over the vertical, which strongly limits its use in studying air quality when we need to know pollutant concentrations close to the surface. Further, satellite measurements are potentially hindered by cloud cover or by large surface albedo (especially for the inversion of aerosols). Finally, if we wish to finely characterize a pollution event in a particular urbanized zone, their spatial resolution ($\approx$ 10 km at best) and temporal resolution (one or two measurements per day) are still generally too low. To clearly understand a pollution event over a duration of a few days, different methods are implemented to make use of these data. We can, first of all, prioritize the analysis of events by taking chemical species with a long lifetime and a clear spectral signature. We can also combine different sensors in order to reduce the uncertainty associated with each. The study can also consist of studying trends: the lack of spatial and temporal coverage is thus compensated through the accumulation of data and their statistical use. An example of a measurement/model comparison is depicted in Figure 7.3.

Satellite	Sounder	Instrument	Pollutant
Nasa/Terra (1999)	Mopitt (TIR + NIR)	Correlation spectrometer. Pixel: 22 km Revisit 3 days	CO
Nasa/Terra (1999)	Modis Visible	Imaging spectrometer. Pixel: 250m Swath 2330 km	Aerosol: type, size, optical prop., AOD
Nasa/Aqua (2002)	Airs (TIR)	Network spectrometer. Pixel: 13.5 km Swath: 1650 km	CO, O_3, NH_3
Nasa/Aqua (2002)	Modis (Visible)	Imaging spectrometer. Pixel: 250-1000 m Swath: 2330 km	Aerosol: type, size, optical properties, AOD.
Nasa/Aura (2004)	OMI	Spectrometer (UV-vis) Pixel: 13 km Swath: 2600 km	NO_2, O_3, SO_2, AOD
Nasa/Aura (2004)	TES (TIR)	Fourier transform spectrometer. Pixel: 5.3 km	CO, O_3, NH_3, SO_2
Cnes/ Nasa Calipso (2006)	Caliop + IRR	Lidar (polarization orthogonal) + Imaging	Vertical profiles, aerosols and clouds
Cnes/ Eumetsat/ Metop	Gome-2 (UV-vis)	UV-vis spectrometer Pixel: 40 km Swath: 1920 km	NO_2, O_3, SO_2, CH_2O
Cnes/ Eumetsat/ Metop	Iasi (TIR)	TF spectrometer. Pixel: 12 km Swath: 2400 km	CO, O_3, NH_3, SO_2, CH_3OC, HCOOH
Cnes/Parasol (2005-2013)	Polder	Imaging radiometer Pixel: 6 km Swath: 1600 km	Aerosol: type, AOD, size, optical properties
Sentinel 5 (2018)	Tropomi (UV, UVIS, NIR)	Spectrometer Pixel: 3km Swath: 2600km	AOD, ANG, O_3, NO_2, aerosols

Table 2.2. *Some examples of satellites used to analyze atmospheric composition. Notations: TIR - thermal IR; NIR - near IR; VIS - visible; UV - ultraviolet*

2.2.3. *Micro-sensors*

For the past few years, there have been new means of measurement, called "micro-sensors" *(Chatzidiakou et al. 2019)*. They are developed and used to address the problem of the impact of pollution on health: to refine current studies, it is necessary to be able to measure human exposure to the closest scale of interest: the individual. These sensors can therefore be installed at a fixed place or carried (and therefore transported) by an individual. At present, due to its compactness, the technology embedded in the sensor is rudimentary and therefore we have measurements with low precision **(Moltchanov et al. 2015)**, **(Fishbain et al. 2017)**. After the data are processed (bias removal, noise-filtration, etc.), the challenge is to compensate for the sensor's lack of precision by increasing the number of sensors used, as they are very cheap. It is thus possible to consider using them to calculate spatial gradients rather than absolute values. The next challenge is to compare data with high spatial resolution with these data. There are many acquired measurements and these have led to the rise of a new discipline in managing "big data" **(Reis et al. 2015)**.

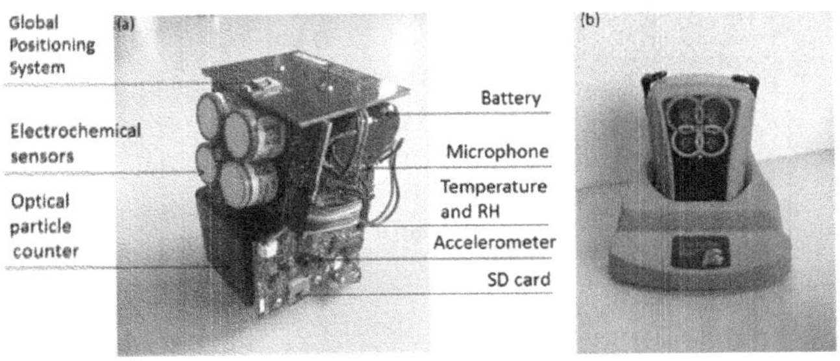

Figure 2.7. *Example of micro-sensors developed to measure the pollution experienced by an individual, and consequently to estimate the exposure of a population. Photo from Chatzidiakou et al. (2019). For a color version of this figure, see: http://www.iste.co.uk/menut/modeling.zip*

2.3. Measurement campaigns

There are many campaigns to measure regional atmospheric pollution and Table 2.3 lists only a few examples. There is sometimes a "special issue" for a project that readers can access, for example:

– MEGAPOLI: https://acp.copernicus.org/articles/special_issue248.html.

– CHARMEX: https://acp.copernicus.org/articles/special_issue334.html.

– DACCIWA: https://acp.copernicus.org/articles/special_issue914.html.

These campaigns can be focused on a single city (usually a megacity, that is, with around 10 million inhabitants at least) or can focus on an entire region. Depending on their spatial coverage, the research theme allows them to cover various subjects, from urban exposure to long-distance transport. Measurement campaigns can last from between a few days to a few weeks. They are extremely costly and therefore quite rare. They mobilize many researchers and are generally carried out within the framework of an international research project. For example, European projects like FP6 and then FP7. They are dedicated to understanding a process (e.g. turbulence in the stable layer, secondary organic aerosoles). They often combine simultaneous and colocalized measurements of meteorological and chemical parameters. These campaigns try to associate the maximum number of different instruments to "close" a problem. The instruments used are not necessarily "routine" instruments, but are often laboratory prototypes. These instruments are intended to measure more variables with greater precision. They require operators and the measurements are not continuous.

Name	Main theme	Reference	Year
ESQUIF	Pollution in the Paris region	**(Menut et al. 2000)**	2000
ESCOMPTE	Pollution in the Marseille region	**(Cros et al. 2004)**	2001
TROCCINOX	Tropical convection and NO_x	*(Huntrieser et al. 2008)*	2005
FUMAPEX	Urban pollution in Europe	*(Baklanov 2006)*	2006
AMMA	West African monsoon	**(Lebel et al. 2010)**	2006
REPARTEE	Pollution in London	*(Harrison et al. 2012)*	2006
MILAGRO	Pollution in Mexico	*(Molina et al. 2010)*	2006
MEGAPOLI	Pollution in the Paris region	*(Beekmann et al. 2015)*	2008
EUCARII	Aerosols in Europe	*(Kulmala et al. 2011)*	2008
ARCTAS	Pollution in the Arctic	*(Jacob et al. 2010)*	2008
CHARMEX	Pollution in the Mediterranean region	**(Dulac et al. 2013)**	2013
DACCIWA	Aerosols/radiation in the Gulf of Guinea	**(Knippertz et al. 2015)**	2018

Table 2.3. *A few examples of measurement campaigns related to the study of atmospheric composition for air quality*

Measurement campaigns like this generally unfold within a research project and require many distinct steps. The whole process takes a few years.

– A researcher or team of researchers identifies a scientific question that seems crucial enough to justify raising a large research budget. Bodies that are capable of coordinating and financing projects of this type are the European Union, for example, through projects such as FP7, H2020, etc. This group will coordinate the project and choose an individual representative to be the project coordinator.

– The coordinator then creates a consortium, that is they contact and bring together various researchers, with the aim of working across all specialities required for the project. For atmospheric pollution, this means bringing together specialists in instrumentation and modeling, in physics and atmospheric chemistry.

– The project takes shape through a "white paper" that lists all of the scientific questions and obstacles to be overcome, and specifies the methodology and a schedule.

– If the project is accepted and financed, the "measurement campaign" will be implemented. Instrumentation researchers will deploy their instruments (surface and airborne measurements) and await the occurrence of the meteorological and pollution events associated with their scientific question. They may be helped by modelers who will make predictions up to a few days ahead in order to help them make a decision about an *intensive observation period* (IOP). This is the most constraining period in the project: (i) instruments must be set up across the zone of study, often abroad and with fewer facilities than in the coordinating laboratory; (ii) it is essential to stay on-site during the entire campaign and await favorable situations; (iii) the team must know how to handle breakdowns and other urgent problems during the campaign.

– After acquiring the data, "quality control" processing must be carried out in order to create a database, available to all members of the project.

– The measurement teams analyze all the data and respond to certain scientific questions based on these results. The modeling teams analyze the documented situations, comparing the results from the models with the measurements, and may respond to additional scientific questions. If the team can answer these questions only with measurements, then it is all the more important to support these with comparisons with the measurements to ensure relevance.

– The full set of results may be published in peer-reviewed journals and conference presentations.

An example of a comparison between a model and data from the CHARMEX campaign is presented in Figure 7.4.

2.4. Databases

Irrespective of whether the measurements are routinely taken or are part of a measurement campaign, the data are made permanently available by being entered into a database. These bases have evolved enormously in recent years, especially with new regulations that require them to be made publicly and easily accessible. They now take the form of open websites. Table 2.4 presents databases that can be used for pollution studies. There are three types of databases: (i) databases with geophysical characteristics; (ii) databases with routine measurements (i.e. automation stations or radiosondes); and (iii) databases with measurements from campaigns. These databases, populated by campaign measurements, are generally established with a delay of a few years in order to allow the participants in the projects to first make use of their own measurements.

2.5. Norms and legislations

In France, the law governing atmospheric pollution is LAURE (*Loi sur l'Air et l'Utilisation Rationnelle de l'Energie* – Law on Air and the Rational Use of Energy). This law, dated December 30, 1996 has a framework covering three major themes: rationalizing energy use, defining public policy integrating air as a subject in urban development and the principle that *the right to breathe air that does not harm our health is guaranteed to all.* This law made it mandatory to have air quality monitoring carried out by the government, define quality objectives and inform the public.

Different management tools were suggested to achieve these goals: a Regional Air Quality Plan, Atmosphere Protection Plans and an Urban Displacement Plan (UDP) (for agglomerations with over 100,000 inhabitants).

Name	Main theme	Reference
	Geophysical parameters	
EMEP	Anthropogenic emissions	https://www.ceip.at/webdab-emission-database
ISRIC	Surface and soil	https://www.isric.org
IIASA	Surface and soil	http://webarchive.iiasa.ac.at/Research/LUC
JRC	Surface and soil	https://esdac.jrc.ec.europa.eu
	Routine meAsurements of atmospheric concentrations	
EMEP		https://projects.nilu.no/ccc
EEA	Surface pollutants in Europe	https://www.eea.europa.eu/data-and-maps
AERONET	Optical thickness	https://aeronet.gsfc.nasa.gov
	Measurement campaigns	
NASA	All databases	https://earthdata.nasa.gov

Table 2.4. *A few examples of databases of interest in the study of atmospehric pollution*

A *national air quality body* was set up to implement this law. It comprised three different entities: the Environment Ministry, that is, the Ministry for Ecological Transition and Solidarity (and more specifically, the Air Quality Bureau), the association for air quality monitoring and LCSQA (*Laboratoire Central de Surveillance de la Qualité de l'Air* – Central Laboratory for Air Quality Monitoring). A synthetic index of air quality, the Atmo index, was also defined. It was created by the Atmo France association, which was established in 2000. This association brings together various Air Quality Monitoring Associations (AASQA) in France. It thus groups organizations in charge of measurement. The Atmo index is meant to be very simple and synthetic: the aim is to use a single number and a single color to indicate the air quality in a given region. It is an increasing index from 1 (very good) to 10 (very poor). There is also a European index that includes five different values, representing categories that range from *very low* to *very high* in terms of the average intensity of pollution measured across all stations in an agglomeration. Table 2.5 presents all of the organizations in charge of information on air quality in France and Europe.

The regulation is applicable to six different pollutants: sulfur dioxide (SO_2), ozone (O_3), nitrogen dioxide (NO_2), particles ($PM_{2.5}$ and PM_{10}) and carbon monoxide (CO). Based on the AASQA data, other pollutants are also monitored as they are considered problematic for health, even if the law does not yet consider that they must be covered by mandatory monitoring. We thus define two thresholds: an information threshold and an alert threshold. Each of these corresponds to air concentrations with thresholds. As there are many surface measurement stations for each of the pollutants, the thresholds are established in different ways. For ozone, nitrogen dioxide and sulfur dioxide, there is information or an alert if the threshold is simultaneously exceeded across three measurement stations in the monitoring network, including at least one background station. The information thresholds defined and used in France are presented in Table 2.6.

Acronym	Complete name and website	Information
ADEME	*Agence de l'Environnement et de la maitrise de l'energie* (Environment and Energy Control Agency) https://www.ademe.fr/expertises/air-bruit	Documentation
ATMO	Atmo France https://atmo-france.org/	Documentation
MTES	*Ministèere de la transition écologique et solidaire* (Ministry for Ecological Transition and Solidarity) https://www.ecologique-solidaire.gouv.fr/	Documentation
PREVAIR	*Prévision nationale de la qualité de l'air* http://www2.prevair.org/	Modeling (national air quality prediction)
EAQI	European Air Quality Index https://www.eea.europa.eu/themes/air/air-quality-index	Measurements
AirQualityNow	Air Quality in Europe http://airqualitynow.eu/	Measurements

Table 2.5. *Organizations in charge of information on air quality*

Pollutant	Information threshold (mean)			Alert
	Year	Day	Time	
Ozone (O_3)			$180 \mu g/m^3$ (over 8 hrs)	$240 \mu g/m^3$
NO_2	$40\ \mu g/m^3$		$200\ \mu g/m^3$	$400\ \mu g/m^3$ (over 3 hrs)
SO_2		$125 \mu g/m^3$	$350 \mu g/m^3$	$500 \mu g/m^3$ (over 3 hrs)
CO			$10\,000\ \mu g/m^3$	
Lead	$0.5\ \mu g/m^3$			
Benzene (C_2H_6)	$5\ \mu g/m^3$			
PM_{10}	$40 \mu g/m^3$	$50 \mu g/m^3$ (35 days/year)		
$PM_{2.5}$		$25 \mu g/m^3$		

Table 2.6. *Information and alert thresholds*

Table 2.7 presents a synthetic air quality balance and how it has evolved over the past 20 years.

Compound	Evolution	Characteristics
Sulfur dioxide	Improvement (strong)	Peaks in winter (industrial sites)
Nitrogen dioxide	Improvement (moderate)	Peaks in winter (threshold not reached)
Particles	Improvement (very moderate)	(Threshold not reached)
Ozone	Degradation	High background levels, peaks in summer
Carbon dioxide	Improvement (heavy)	

Table 2.7. *General trends in surface pollutant concentrations in Europe between 1995–2015*

3

General Principle Behind Modeling and Its Application to Meteorology

There are many processes involved in the evolution of atmospheric composition on a regional scale. One way of taking into account all these processes is to develop **integrated numerical modeling tools**. This chapter first introduces the basic principles of the modeling used for atmospheric pollution. As the representation of meteorological conditions is the primary prerequisite for modeling atmospheric pollution, the section explaining what a model is will be based on meteorological modeling.

3.1. What is a model?

3.1.1. *Definition of a numerical model*

A numerical model is a set of equations that describe a certain physical reality. Only a *certain* physical reality, because the physical processes that have actually been observed are generally too complex to be able to be described in their integrity, and above all, too complex to be calculated. A model is therefore a **mathematical tool** whose only objective is to aid in the comprehension of a complex physical phenomenon. In order for this understanding, which is called the *analysis*, to be as faithful as possible, the set of equations in the model must be adapted *as best as possible* to the problem being treated. There is, therefore, no unique physical model. Instead, there are almost as many models as there are specific problems to be treated. We can go even further, saying that for the same type of problem there exist many different models, that is, many different ways of simplifying the same physical phenomenon being studied. The model consists of equations, which are themselves made up of variables. The fact that a limited number of equations are chosen, and therefore that there are a limited number of associated variables, makes this a **discrete system**. **Discretization** is when these equations are decomposed in time or space in order to integrate them. In meteorological and atmospheric chemistry case studies, the model is thus made up of a set of equations describing the atmosphere.

For meteorology, these equations are based on the **Navier–Stokes equations**. These equations include the spatial and temporal evolution of wind, temperature, humidity and air pressure (**Pielke 2013**). The major difficulty is that all these variables are **inter-dependent**, which makes this a **nonlinear system**. In this case, we have a problem that is very sensitive to the initial conditions. On the other hand, and for chemistry transport, the equations to be solved consist of the chemical mechanism. If the chemical species are also interdependent, then the system is again nonlinear, but much less than in meteorology. We then have a **problem with boundary conditions**. These boundary conditions are principally at the surface, with emissions of pollutants being the major cause for the variability of the results.

There are many interests that lead to the development and use of a model. The first is a desire to **understand** complex physical phenomena. How do large climatic phenomena, such as the El Nino, occur? What impact does a volcanic eruption have on the atmosphere? How do bush fires contribute to eventual climate change? How is atmospheric pollution transported from a city to its outskirts? etc. Another significant advantage that models offer is being able to **reconstitute** inaccessible information. First of all, in time, where we can reconstitute an environment from the past or carry out climatic studies on warming. Then, in space, understanding a physical phenomenon requires measurements. However, in practice it has been seen that it is impossible to carry out these measurements everywhere and at any given time. Only a modeling tool can offer regular "fields" for the variable we wish to study. Finally, we cannot, and do not know how to, measure all quantities that can come into play in the complex system of dynamics: a model is thus a great help in "reconstituting" unmeasured variables. A model also makes it possible to **implement scenarios** related to the evolution of the system. For example, in regional pollution models, we can vary anthropogenic emissions and study what effect this has on concentrations. Finally, a model makes it possible to predict events, and today it has become an essential tool in air quality management.

The models used may be deterministic or statistical. If we want to study the transport of a mass of air and remain focused on this mass, the model of choice would be a Lagrangian model. However, if we want to study a region where air masses are in transit, with a focus on the region, it would be preferable to use a Eulerian model. Depending on the scale of the process being studied, we can use either a global model, a regional model or a highly local model (e.g. a neighborhood in a city). In the context of this document, and with the aim of studying polluted regions with accuracy and the possibility of simulating varied situations, our model of choice will be a **deterministic, regional Eulerian model**.

3.1.2. *Journals on modeling*

There are many journals that discuss geophysical modeling. Table 3.1 lists a few of these and indicates whether or not the articles are *open access*. The term "depending on the authors" denotes the fact that for a few years now it has been possible to publish an open-access article (generally by paying an additional amount, with respect to the initial price of the article). This is a recent possibility and the oldest articles cited here are often behind a paywall.

Journal	Open access
Atmosphere	Yes
www.mdpi.com/journal/atmosphere	
Atmospheric Chemistry and Physics	Yes
https://acp.copernicus.org/	
Atmospheric Environment	Depending on the authors
https://www.journals.elsevier.com/atmospheric-environment	
Atmospheric Research	Depending on the authors
https://www.journals.elsevier.com/atmospheric-research	
Boundary Layer Meteorology	Depending on the authors
https://www.springer.com/journal/10546	
Bulletin of the American Meteorological Society	Depending on the authors
https://journals.ametsoc.org/view/journals/bams/bams-overview.xml	
Environmental Modelling and Software	Depending on the authors
https://www.journals.elsevier.com/environmental-modelling-and-software	
Geoscientific Model Development	Yes
https://gmd.copernicus.org/	
Journal of the Atmospheric Sciences	Depending on the authors
https://www.ametsoc.org/index.cfm/ams/publications/journals/journal-of-the-atmospheric-sciences/	
Journal of Geophysical Research – Atmospheres	Depending on the authors
https://agupubs.onlinelibrary.wiley.com/journal/21698996	
Monthly Weather Review	Depending on the authors
https://www.ametsoc.org/index.cfm/ams/publications/journals/monthly-weather-review/	

Table 3.1. *Examples of scientific journals on atmospheric modeling*

3.1.3. *Different types of Eulerian models*

Eulerian models (Figure 3.1) study processes in a fixed domain: the "mesh" or the "grid". A parcel of air can enter or exit the domain, and the primary objective of the simulation is to evaluate what happens to a particular fixed point in space over time. In this family of models, we find different configurations that are dependent on the **number of spatial scales** that we consider: from 0D to 3D, that is, from considering only time (t) to considering all dimensions (t,x,y,z).

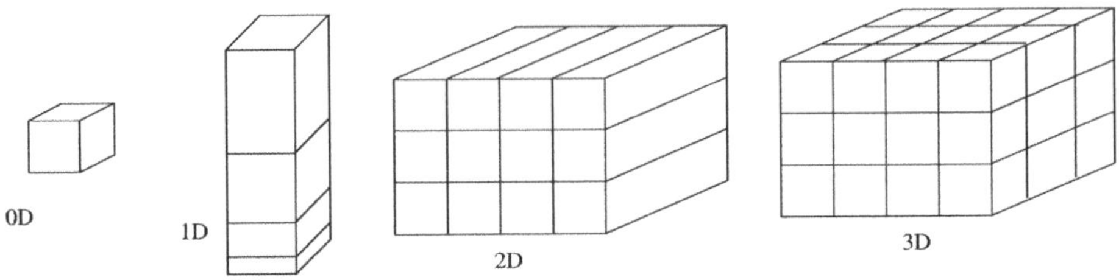

Figure 3.1. *The meshes for 0D, 1D, 2D and 3D Eulerian models*

0D models are primarily "laboratories" to study and perfect parameterizations. By removing all concept of space (quantities are calculated at a single point), these models are used above all to gauge the relevance of a new parameterization. These models can be used in dynamics or in chemistry. For example, the GECKO model *(Aumont et al. 2005)*, makes it possible to simulate complex chemical mechanisms with a realistic computation time. **1D models** are commonly used to quantify the impact of parameterizations on a main flux. The results are **vertical profiles**. While it is light, in numerical terms, this model can only be used when the wind is very weak, that is, when the influence of the flux around the site of study is not a dominant factor affecting the evolution of the variable being studied. **2D models** are used less often than 1D and 3D models. They are generally simplified versions of existing 3D models. The calculation is carried out in a **vertical plane** and one of the two axes (x or y) is given greater importance. The choice is finally imposed by the principal wind flux: the chosen axis being the wind direction. This type of model is used in the case of fluxes that are highly constrained in one direction, for example, a sea-breeze constrained by relief. The advantage of 2D models with respect to 1D models is that we can take into account the relief and the influence of the wind from one horizontal mesh to another. The constraint is that the wind must not change direction for the duration of the simulation. **3D models** represent the most complete version as we work in three dimensions of space. This is also the version that has the highest numerical cost. The advantage of these models with respect to 2D models is that we no longer have to worry about creating a simulation where the wind only fluxes in one direction. These are the most widely used models today: the advances in terms of computation power have led researchers to almost completely turn away from 1D and 2D versions.

3.1.4. *A few current meteorological models*

The general principle is to integrate the Navier–Stokes equations in time and space. This system of equations is nonlinear and consists of calculating, for each time-step, the freshly updated value of pressure, wind, temperature and humidity in the atmosphere. In order to do this, we must calculate (i) the law of perfect gases, and (ii) the equations for the conservation of mass (called the continuity equation), momentum, energy and humidity. The calculations decompose each equation into a principle term and a turbulent fluctuation: these are the Reynolds equations. The system is quite complex and for further details, you can find perfectly described information in **Pielke (2013)**. Many models exist and are available as *open-source* models. They are being continuously developed and are supplied with all the databases required to use them *(Baklanov et al. 2014)*.

Model	Spatial	Characteristics	Reference
MPAS	Global	Model for Prediction Accross Scales http://mpas-dev.github.io/	**(Skamarock et al. 2012)**
GFS	Global	Global Forecast System	**(Kalnay et al. 1996)**
IFS	Global	Integrated Forecast System ECMWF	**(Morcrette et al. 2009)**
COSMO	Regional	Convective-scale numerical weather prediction	**(Baldauf et al. 2011)**
WRF	Regional	Weather Research and Forecasting Model http://www2.mmm.ucar.edu/wrf/users/	**(Powers et al. 2017)**
Meso-NH	Regional	Non-hydrostatic meso-scale smodel	*(Lac et al. 2018)*

Table 3.2. *Examples of global and regional meteorological models*

Table 3.2 presents a non-exhaustive list of these models. The codes for GFS, IFS and global and operational models are not freely accessible, and are only used by the operational centers that develop them. On the contrary, the meteorological fields that are calculated are made available to the community: the IFS model from the European Center (ECMWF) is used only for academic research, while the American GFS model from the NCEP is entirely freely accessible. Among the regional models, the most widely used is the WRF model. Made available through open-source code from inception, it has since then gathered a large community of users and contributors. Today, it is a veritable toolbox, offering the choice of various parameterizations for each process.

3.2. Constraints on a regional model

3.2.1. *Definition of the domain of calculation*

The mesh is the main calculation support and also the final representation that we will have of the results. We must remember that the results obtained by a model are only the **average values in each grid cell**. The choice of mesh is, above all, a **compromise** between different constraints:

– **In the domain:** simulating a plain or a mountainous region does not produce the same mesh. The greater the relief, the larger the chosen horizontal resolution, in order to avoid "smoothing" the relief too much and thus influencing the atmospheric flux.

– **At the edges of the domain:** the boundaries of the domain must be chosen carefully so as not to "cut across" a process that is important for the results. In meteorology, we will avoid placing a domain boundary on an important relief feature. For chemistry transport, we will avoid placing the boundary on an important source (in the middle of a megacity, for example).

– **Data available for validation:** if we have a limited amount of data for validating the results of the model through comparison with measurements, it may be useful to consider using a mesh based on the representativity of the data.

– **Computing power:** all the above points can push us toward finer and finer meshes: the available computing power is the constraint that will limit this trend.

– **Turbulence:** we can move toward finer meshes as long as turbulent phenomena are well represented. For most meso-scale models, turbulent processes that are considered to be "subgrid" processes are parameterized. If we wish to have a resolution smaller than a kilometer, the model must be changed: turbulence must be explicitely resolved to realistically take into account eddies. We then use a model called a *large Eddy simulation.*

If the complete domain is too small, then the simulation is no longer as representative of the complete environment that we are trying to simulate. However, if it is too large, with respect to the minimum accuracy required, we risk having a prohibitive computation domain (including an enormous amount of output data). In parallel, the mesh resolution must be adapted to the size of the domain and to the process being modeled. If the resolution is too low, the obtained results are only the mean values of very large zones, which would be more contrasted than what we were trying to reproduce: we may "miss out" on processes such as breezes, for example. However, if the resolution is too high, the computation time becomes unnecessarily prohibitive. The **time step** must be chosen in direct relation to the choice of mesh, and more generally, with respect to the process that we wish to model. This time step corresponds to the temporal increment between two successive integrations in the model: from t to $t + \Delta t$. We must ensure that the grid is appropriate to the flux that is being studied (the site and period) and therefore based on the wind velocity being considered across this period. This is ensured by calculating the Courant–Friedrich–Levy number (CFL), such that:

$$CFL = |U|.\frac{\Delta t}{\Delta x} \qquad [3.1]$$

Throughout the simulation, we must have: CFL $\approx$ 1. If this is not the case and if CFL $<<$ 1: the distance covered during the Δt due to the advection is smaller than Δx and the transport schemes will generate an artificial accumulation of kinetic energy in a mesh. If CFL $>>$ 1, the distance covered during Δt due to advection is greater than Δx: some grid cells are avoided. The simulation may diverge or, if it does not diverge, the results are incorrect. In current models, these criteria are automatically calculated and the time step is continuously adjusted during the simulation.

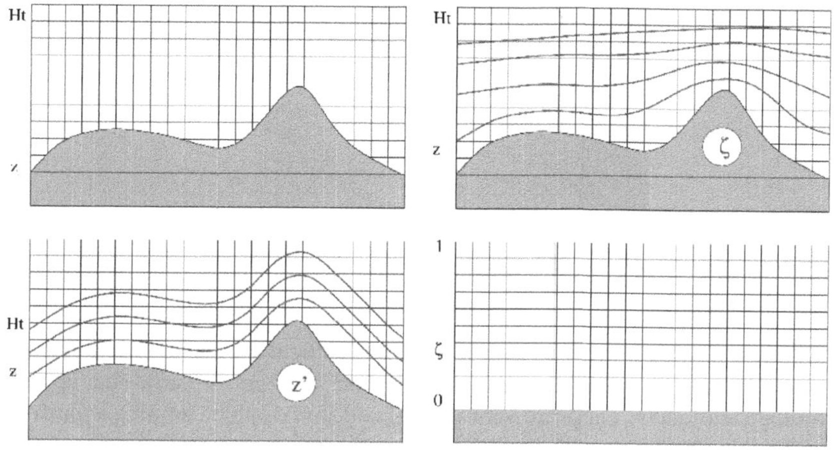

Figure 3.2. *Examples of vertical meshes, with or without relief and with different changes in coordinates. For a color version of this figure, see: http://www.iste.co.uk/menut/modeling.zip*

The **vertical mesh** is as important as the horizontal mesh. It may be irregular that the variables representing this dimension may be used differently in the equations. In the case of regional modeling of the boundary layer, the vertical mesh is chosen to be accurate close to the surface (≈10–20 m of thickness for the first mesh) and then relaxed with altitude. The dynamic equations, as they are resolved, may undergo a change in reference frame in order to accurately account for irregularities due to relief. For example, the vertical coordinates to take into account the relief are defined by the coordinates ζ: the relief is characterized by its height Z above the ground: $Z = h(x,y)$. The highest point of the domain is H_t. The relief can simply be removed: $z' = z - h(x,y)$. The coordinates are changed, with ζ varying between 0 and 1: $\zeta = (z - h(x,y))/(H_t - h(x,y))$. To best simulate the boundary layer and the free troposphere, meteorological models use hybrid coordinates called "σ-pressure" (*sigma-pressure*) coordinates (**Jacobson 1999**).

$$\sigma = \frac{p - p_t}{p_s - p_t} \qquad [3.2]$$

with p being the atmospheric pressure at any point, p_s being the surface pressure and p_t being the pressure at the top of the domain. At the surface, σ = 1, at the top of the domain σ = 0. Table 3.3 presents an example of a vertical mesh expressed in σ-pressure coordinates.

Level	σ	Pressure (hPa)	Altitude (km)
1	0.995	995	0.04
2	0.97	972	0.22
3	0.90	905	0.75
4	0.75	763	2.0
5	0.60	620	3.6
6	0.45	478	5.5
7	0.30	355	8.2
8	0.15	193	12.4
9	0	50	22.5

Table 3.3. *Example of σ-pressure coordinates for a model with nine vertical levels*

Present-day models are evolving toward non-structured meshes. The techniques that already exist in CFD (*computational fluid dynamics*) could be applied to meteorological models. There are various applications possible: refining a mesh where there are strong horizontal gradients, like in a coastal or mountainous region, or, on the contrary, relaxing it toward boundary conditions. The MPAS model **(Skamarock et al. 2012)** uses a mesh that is unstructured on a global level. Applications to chemistry-transport models are presented in *Garcia-Menendez and Odman (2011)*. However, there are currently few regional models that have unstructured meshes. One of the few such models has been developed by *Zheng et al. (2015)* and an example is given in Figure 3.3.

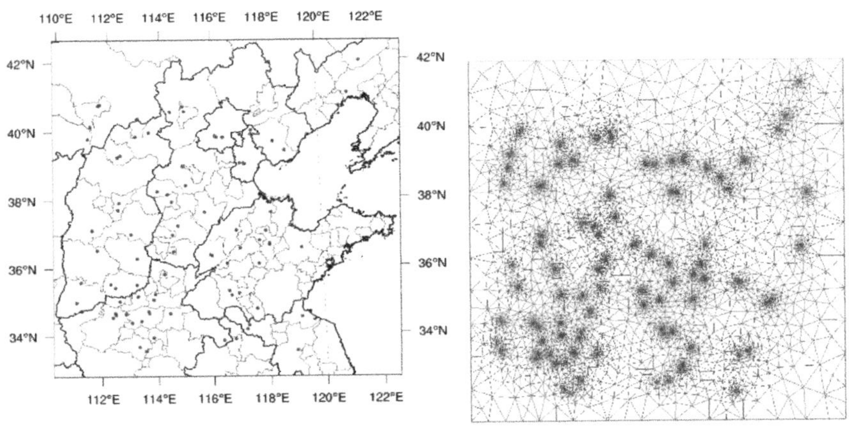

Figure 3.3. *Example of an unstructured mesh currently being developed to model chemistry transport. The map depicts factories and the mesh is mainly refined around these points of emission. Adopted from* Zheng et al. (2015). *For a color version of this figure, see: http://www.iste.co.uk/menut/modeling.zip*

The choice of the horizontal and vertical spread of the domain and the horizontal and vertical resolution will determine the accuracy of the results and the computation time. For each study, these choices can be reviewed based on the constraints with respect to simulation time, the lifetime of the process being studied, etc.

3.2.2. *Initial conditions and boundary conditions*

In contrast to global modeling, the most limiting constraints in regional modeling are the **boundary conditions**. This means that since the simulation domain is of limited area and masses of air are transported, we must be able to provide information on the boundaries of the domain to know what will enter and exit during the simulation. The less a domain is subject to these boundary conditions, the more representative the results within the chosen domain will be. This thus limits the size of the domain: it is always better to have a domain that is larger than required. We must also avoid defining a domain whose boundaries pass through *obstacles* to calculations. These obstacles may be regions with high thermal gradients or mountainous regions. It is known that meteorological calculations are tricky here and therefore we must not mix up the model computations (within the domain) and the boundary conditions (just outside the domain).

Boundary conditions must be implemented with certain constraints. First of all, there is the potential problem of **different parameterizations**. If the two models (that of the forcing and the regional model) have not used the same parameterizations to calculate the same variable, there may be biases in the regional computation. The problem of the representativity of the ranges: the two models with different scales will present variables that do not represent the same environment. This is the problem with the average on the borders: based on the regional mesh, we must see how to optimally interpolate (or average) the low resolution ranges to a higher resolution. Current meteorological models are able to manage this problem in that they use the same tool for the embedded models: these are *nesting* techniques.

As far as possible, the initial and boundary conditions must be available for all the variables that are then calculated and/or are inter-dependent in the model: the zonal and meridional wind components, u and v, the temperature T, the pressure P and the specific humidity q. Knowing that dynamic models are strongly nonlinear, the **initial conditions** must be as precise as possible. There are two possible methods. We can initialize a regional model based on the ranges on the largest scales (taken from global models). If we do not have these data, it is possible to run the simulation for a long time before the case study, so that the ranges are balanced: this is the "spin-up" problem. In practice, in order for the spin-up time to be correct, we estimate the time required for a mass of air to travel through the domain of study at its largest width and with a medium wind. For example, for a simulation of Europe, in order to not miss out on important events, a simulation must begin 2–3 weeks before the date that we want to begin analyzing the results. In the specific case of forecast simulations, we use the previous simulation to start the new prediction. We carry out a simulation which starts by using the last range calculated for the previous simulation. The advantage is that the ranges calculated correspond exactly to the domain that is being used. We are clearly only shifting the problem.

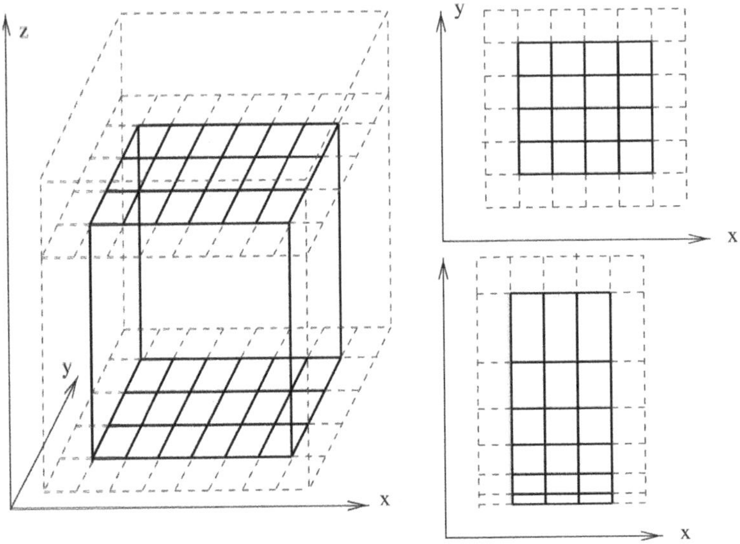

Figure 3.4. *An example of a mesh that includes boundary conditions. For a domain with limited area, we must provide the values for variables for the edges and the summit of the domain throughout the simulation. On a large scale, a meteorological forcing every 3 or 6 h may suffice; on a regional scale, taking hourly model data are recommended. For a color version of this figure, see: http://www.iste.co.uk/menut/modeling.zip*

Since the meteorological system is strongly nonlinear, the initial conditions are primordial: each error may be amplified and propagate through the domain. The boundary conditions in a regional model are important for the quality of the results. Depending on the chosen data source, and its spatial and temporal resolution, it is possible to create a bias in the results which will combine with other errors in the model and thus be difficult to interpret.

3.2.3. *Surface data*

The surface is one form of boundary conditions. Depending on the parameters to be established, these data may be static or dynamic, i.e. variable in time. Current models are generally provided with their own databases for surface data, often with different resolutions possible in order to be able to follow the user's mesh fineness. In general, a meteorological model will require: data on relief, surface type (*landuse*), albedo, emissivity, roughness length, vegetation and soil type. Table 3.4 offers examples of values to understand the orders of magnitude.

Soil type	Water	Forest	Bare mineral	Diffuse construction	Mixed construction	Dense construction
Albedo: A	0.08	0.16	0.25	0.18	0.18	0.18
Emissivity: ϵ	0.98	0.95	0.965	0.88	0.88	0.88
Thermal inertia of the soil: C_{sol}	7.6	11.0	5.0	7.5	5.7	3.9
Sensible urban heat flux: Q_0	0	0	0	10	15	30

Table 3.4. *Orders of magnitudes of meteorological parameters depending on the type of surface*

Among the required parameters, roughness length (z_0) remains the most difficult to establish as it does not directly correspond to any measurable quantity. We define three different roughness lengths: (1) dynamic: z_{0m}, (2) thermal: z_{0t} and (3) humidity: z_{0q}. In practice, there is very little information about these quantities, but they have a major impact on turbulence in the surface layer and, consequently, a major impact on the entire development of the ABL during a convective period. For a first approximation, it is not uncommon to use: $z_{0t} = z_{0m}.e^{-2}$ and $z_{0q} = z_{0t}$. The only remaining unknown is then z_{0m}. "Typical" values are z_{0m}=0.1 m in a rural environment and z_{0m}=1 m in the city. This length symbolizes the capacity of the surface to slow down air flux, that is, the height above ground where we have $\overline{u} = 0$ at $z = z_{0m}$.

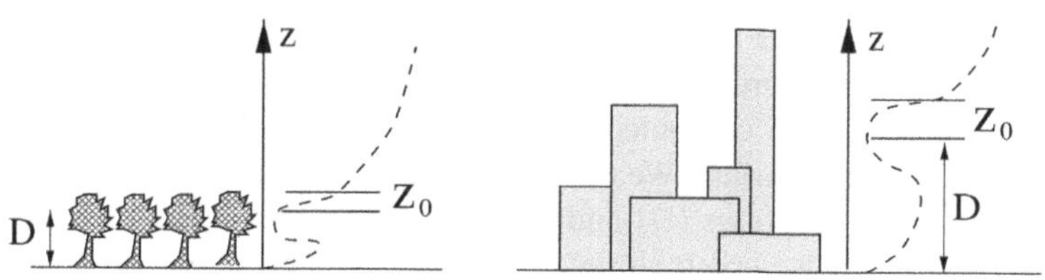

Figure 3.5. *Schematic illustration of height of displacement depending on the environment*

As seen in Figure 3.5, this concept is supplemented by the concept of displacement height: when the vegetation is dense enough so that wind flux cannot penetrate here, the neutral wind profile is no longer logarithmic from the ground, but from a height h_D. This phenomenon may also be found with urban buildings if there is a high spatial density. Table 3.5 presents a few orders of magnitude of roughness lengths.

Surface type	z_0 [m]	Surface type	z_0 [m]	Surface type	z_0 [m]
Ice	10^{-5}	Desert	10^{-3}	Forest	1
Snow	10^{-3}	Grass	5.10^{-2}	City (suburbs)	2
Calm sea	10^{-3}	Coppice	0.1	City (center)	5–10

Table 3.5. *Typical values of dynamic roughness lengths z_{0m}*

Surface data defines the entire environment, which will enable the computation of meteorology in the boundary layer. Current models have very comprehensive databases, often on a global scale and with kilometer resolution. However, this does not prevent uncertainty, especially for "derived" parameters, such as roughness length or soil composition. An error in surface conditions will create a constant bias in simulation, which is often difficult to understand and identify.

3.2.4. *Nudging and nesting*

Figure 3.6 presents the two techniques of nesting and nudging. **Nesting** consists of running the same model but with different resolutions and over different computation domains. This is the same as having ideal boundary conditions: the computation schemes are the same and we have a forcing with the desired temporal resolution. **One-way nesting** represents information traveling only from the large domain (D1, in black) to the smaller domain (D2, in red). The information from the larger mesh and at the boundaries of the small domain, D2, will then either be re-copied or interpolated to pass on the information. **Two-way nesting** means that the information travels in both ways. In this case, the high-resolution D2 will be interpolated or averaged to send back information toward the larger domain D1. This approach is very useful when we have to simulate events where there is atmospheric recirculation. **Nudging** is a method for integrating information computed on the larger domain D1 into the domain D2 (and not only at the boundaries). We therefore have grid-points in the smaller domain where there are two data points (D1 and D2). By constraining D2 to follow the trend of the larger scale D1, we come closer to data assimilation. **Spectral nudging** was developed in order to respect the different scales of flux by taking into account only the large-scale forcing and allowing the smaller scale data to be expressed in D2. This makes it possible to establish a cut-off wavelength and only consider forcings for which the wavelength is greater than the cut-off value (**von Storch and Zwiers 2001**).

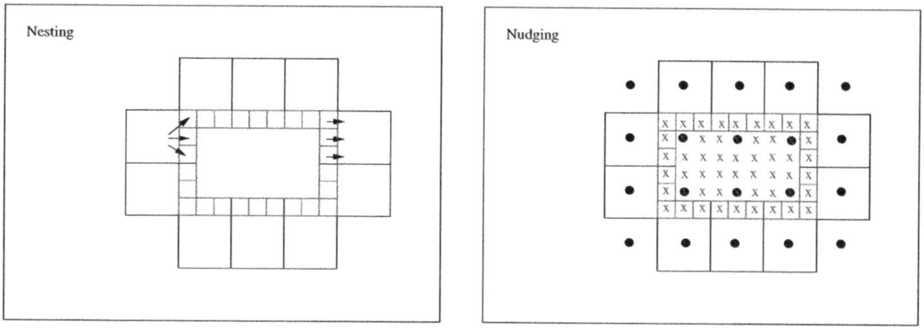

Figure 3.6. *Schematic illustration of nesting and nudging. For a color version of this figure, see: http://www.iste.co.uk/menut/modeling.zip*

Nesting and nudging contain adjustable parameters in models: how they are adapted to the case being studied is a significant constraint on the results. For nudging, we must be able to contain the smaller domain in altitude using the large-scale fields, however we must also leave the model that has better resolution in the boundary layer free so that the turbulence can be optimally calculated.

3.3. Dynamical, physical and parameterizations

A meteorological model will solve Navier–Stokes equations, but in a discrete form, known as Reynolds equations. The Reynolds decomposition consists of splitting a variable c into two: its mean part, $\overline{c}$, and its fluctuation, c' (**Jacobson 1999**): $c = \overline{c} + c'$. The term c' corresponds to the terms fluctuating around the mean value that cannot be explicitly calculated. It is obvious that the greater the resolution of the model, the more we have $c \approx \overline{c}$ and $c' \to 0$. Therefore, if we had enough computing power to carry out a simulation on a global scale with a resolution of 1 mm, we would not require parameterizations. However, this is not the case. We will therefore estimate the fluctuation terms c' by developing the appropriate parameterizations (**Stensrud 2011**). For meteorological models, the calculation of the evolution of $\overline{c}$ is called *dynamical*, and the calculation through parameterization of c' is called *physical*. Parameterization is thus defined as **the estimation, for a certain variable, of its fluctuation, which cannot be explicitly calculated**. This may also be the complete determination of a variable and its fluctuation when it is subgrid with respect to the simulated domain. We will thus seek to diagnose it. This is done by estimating it using a mixture of known terms and parameters. These parameters are coefficients or supplementary equations. Parameterization must respect some principles. First of all, the parameter must be of the same dimension as the unknown. It must also have the same tensor properties (vector, matrix etc.). It must be invariant with respect to the coordinate system or to some other type of transformation on which it is not dependent. Finally, it must respond to the same equations and constraints as the unknown that is being replaced.

3.3.1. *Choosing the timestep*

The choice of the timestep for the calculation will also be the choice for the separation between the mean value and its fluctuation. For example, wind velocity is not a stationary phenomenon: $u = \overline{u} + u'$. Figure 3.7 depicts an example of the recorded values for the wind speed at a given point, based on time. If we average all the values over the entire time interval, we obtain the value depicted by the dotted line, $\overline{u}$. Having values for u that are both smaller than and greater than $\overline{u}$, the balance equation, upon integration, will schematically be $\overline{u'} = 0$ here. We arrive again at one of the definitions of turbulence: the average of all fluctuations is zero. However, if the integration interval is very large, and given that different variables (p, u, v, T) interact, we risk underestimating turbulent processes and only seeing its average value in the system to be resolved. It can therefore be seen that choosing an interval without knowing even the nature of the variability of what we are studying can lead to erroneous results. This is shown in greater detail in Figure 3.7 and if we take the average over the interval t_1: we take the average of the positive values of u, we therefore have $u > 0$ and $u' > 0$, which corresponds well with what has been observed. On the contrary, the average will certainly be underestimated with respect to the real phenomenon. If we choose the interval t_2: we surround t_1, however in this case the average will give $u < 0$ and $u' \approx 0$. By increasing the integration interval we artifically reduce the turbulence and change the sign of the variable being studied. Finally, if we choose the interval t_3: the wind oscillates weakly around its mean value. In this case, we have $u = \overline{u}$ and $u' = 0$, which is again unrealistic, but less aberrant than the choice of t_2. The choice of the time interval is therefore primordial for an accurate description of the turbulence. This is the same for the spatial averages that we may be lead to carry out. Whether in measurement or modeling, the turbulence problem is a tricky one to solve. Within the boundary layer, the turbulence will materialize in the transport of momentum and heat. Turbulence is therefore a problem of scale and we must define its bounds. For the lower limit: it has been shown that molecular diffusion did not impact what happened to flux in the boundary layer. Indeed, the smaller eddies observed are several orders of magnitude greater in size, i.e., a few centimeters. We can thus accurately describe turbulence by using Navier-Stokes equations without any additional equations. For the upper limit: the larger eddies are of the same order of magnitude as the vertical extension of the boundary layer, therefore of the order of 1 to 2 km.

3.3.2. *Diagnostic and prognostic variables*

A **diagnostic variable** is a variable that can be derived at the time t from other flux variables at this same time t. For example: the potential temperature θ_t is a function of temperature T_t and pressure P_t. A **prognostic variable** is a variable that we will try to predict. This is therefore the value of a variable at time $t + \Delta t$ based on the value of other known variables (including itself) at the preceding time t. For example, in meteorology, using the Reynolds system we can estimate the value of the temperature $T_{t+\Delta t}$ from T_t, P_t, u_t, v_t, q_t, *etc.*

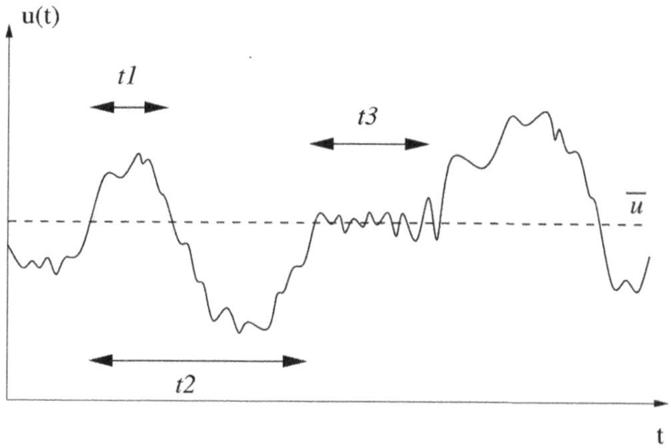

Figure 3.7. *Temporal variation of wind speed values*

For meteorological models, the calculation can be split into two parts: (i) the integration of the continuity equations and equations for the conservation of energy mass and (ii) physical parameterizations that make it possible to "close" the system of these equations by diagnosing turbulent parts or unknown terms. We will then have parameterizations to estimate the processes, such as (i) turbulence in the boundary layer, notably with surface fluxes, (ii) interactions between soil, surface and atmosphere, (iii) parameterizations for the transformation of water in the atmosphere (cloud formation, ice), (iv) convection parameterizations, (v) microphysics and (vi) radiative forcings. As seen in Figure 3.8, the different parameterizations are inter-connected: each of these describe subgrid fluctuations in the atmosphere by involving the same variables. These processes are therefore highly inter-dependent.

3.3.3. *The boundary layer and turbulent fluxes*

The first example is the set of parameterizations that make it possible to calculate surface fluxes and the characteristics of the boundary layer. There are many conceptually different schemes. The knowledge of surface fluxes and convection within the boundary layer will govern the precision of the calculation of meteorology and also of the mixture of pollutants and, therefore, their concentration close to the surface **(Stull 1988)**, see section 1.5. For a mean variable $\overline{C}$, we will have a balance equation of the form:

$$\frac{d\overline{C}}{dt} = \frac{\partial \overline{C}}{\partial t} + \overline{U}\frac{\partial \overline{C}}{\partial x} + \overline{V}\frac{\partial \overline{C}}{\partial y} + \overline{W}\frac{\partial \overline{C}}{\partial z} = \overline{S} - \frac{\partial \overline{u\,c}}{\partial x} - \frac{\partial \overline{v\,c}}{\partial y} - \frac{\partial \overline{w\,c}}{\partial z} \qquad [3.3]$$

with $\overline{S}$ being the source and sink terms affecting what happens to the variable C. The final terms in the equation take the form of a divergence of flux. These terms are said to be "subgrid" and these are the terms that require the development of parameterizations in order to evaluate them. If we consider that conditions are horizontally homogeneous, and if we treat the advection terms separately, the rapid variability of the variable may be summarized by:

$$\frac{\partial \overline{C}}{\partial t} = -\frac{\partial \overline{w\,c}}{\partial z} \qquad [3.4]$$

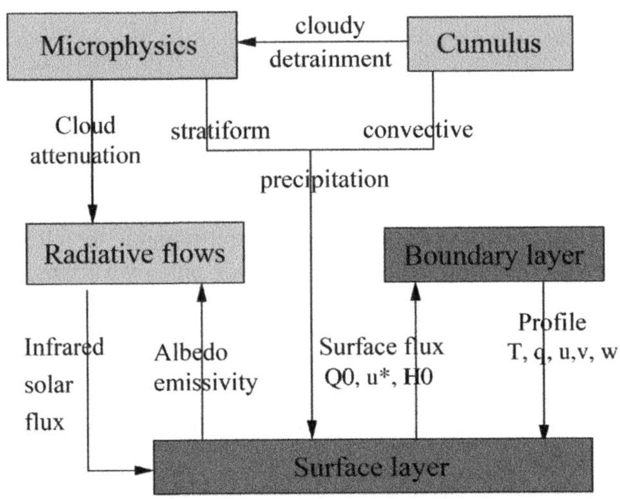

Figure 3.8. *Interactions between the different parameterizations in a meteorological model. For a color version of this figure, see: http://www.iste.co.uk/menut/modeling.zip*

This equation represents the vertical diffusion term for an atmospheric quantity. This shows that the temporal change in a variable linked to turbulence may be estimated by the vertical divergence of the turbulent flux $\overline{w\,c}$. If we consider a convective period and therefore that the turbulence is high and the mixed layer is vertically quite homogeneous, we can integrate the preceding equation over the vertical and we then have:

$$\frac{d\overline{C}}{dt} = \frac{\overline{w\,c_0} - \overline{w\,c_h}}{\overline{h}} \qquad [3.5]$$

with $\overline{w\,c_0}$ being the turbulent flux at the surface and $\overline{w\,c_h}$ being the turbulent flux at the boundary layer top, denoted by $\overline{h}$. For this expression to be evaluated, we must therefore compute the fluxes at the surface and at the top, as well as the height of the boundary layer. Parameterizations were developed accordingly: the flux is expressed as a function of the principal variable itself and as a function of the parameters. This approach is called the *bulk approach*, as the calculation is done "in bulk", considering that the boundary layer is well mixed and, consequently, that the vertical transport is quite homogeneous.

These turbulent terms can be estimated using local and non-local turbulent closures. A local closure is used when an unknown is calculated from values and gradients of known quantities at the same point. A non-local closure is when this same computation uses known values at multiple points in space. For turbulence in the boundary layer, most schemes are first-order turbulent closures. Second-order schemes calculate turbulent kinetic energy (TKE) to diagnose vertical diffusivity **(Mellor and Yamada 1974)**. For a first-order local closure, the theory most commonly used to estimate vertical fluxes in the boundary layer is the K-theory **(Troen and Mahrt 1986)** such that:

$$\overline{wc} = -K_c \frac{\partial \overline{C}}{\partial z} \qquad [3.6]$$

where K represents turbulent diffusivity. We relate a turbulent flux directly to a vertical gradient and to a parameter at the same site. In general, horizontal diffusivity is neglected, considering that the transport schemes used are diffusive enough to represent this process. We focus on the vertical diffusivity (K_z). The dimension of K_z is length times velocity, therefore $m^2.s^{-1}$. In the case of the equation for the conservation of momentum, the equation is the same as finding the flux $\overline{w\,u}$, and therefore of using the vertical gradient of wind. The calculation of this coefficient in a well-mixed layer is generally done using mixing length models. There are many schemes to diagnose the height of the boundary layer **(Shin and Hong 2011)**: the highly local scheme used by **Troen and Mahrt (1986)**, the local "Yonsei University" scheme **(Hong et al. 2006)**, and the non-local scheme from **Pleim (2007)**.

However, at close proximity to the surface, small eddies induce vertical variation in wind and different temperatures **(Deardoff 1972)**. The vertical profile no longer corresponds to a well-mixed layer and there are steep vertical gradients. We must therefore add new parameterizations to describe **surface fluxes in the turbulent surface layer**. In neutral conditions, we have:

$$\frac{dU}{dz} = \frac{u_*}{k\,z} \qquad [3.7]$$

with u_* being the shear velocity, z being the altitude at which we are seeking to establish U and k = 0.41 being the Von Karman constant. The vertical integration of this relationship gives us:

$$U = \frac{u_*}{k} log\left(\frac{z}{z_{0m}}\right) \qquad [3.8]$$

with z_{0m} being the roughness length for the momentum **(Lettau 1969)**. To close this problem, we must therefore determine the shear velocity u_*, the sensible heat flux at the surface Q_0 and the height of the boundary layer $\overline{h}$. In unstable conditions, we must add stability functions to these calculations. In a convective period, we must also add another scale of characteristic velocity: w_*, the vertical convective velocity. We then have the relations:

$$w_* = \left(\frac{k\,g\,\overline{h}\,z}{\rho C_p T}\right)^{1/3} \qquad [3.9]$$

which yield the Monin–Obukhov length L:

$$L = \frac{-zu_*^3}{w_*^3} \qquad [3.10]$$

and the vertical wind gradient close to the surface becomes:

$$\frac{dU}{dz} = \frac{u_*}{k\,z}\phi_M\left(\frac{z}{L}\right) \qquad [3.11]$$

with ϕ_M being the stability function of momentum **(Paulson 1970)**. The same stability functions exist for heat fluxes and humidity fluxes. We thus have a system that can only be iteratively resolved, since the parameters are interdependent. Equivalent “bulk” schemes have been developed to bypass this problem, such as the parameterization given by **Louis et al. (1982)**. These schemes are based on the Richardson layer number, R_{iB}: this expresses the same ratio as the Monin–Obukhov length, that is, the relative weight of the mechanical forcings with respect to the thermal forcings in the mixed layer. However, there is no need for iterative processes to estimate this as it is directly based on the vertical gradients in the principal variables. There are many schemes, along with those discussed previously **(Holtslag et al. 1995; Brutsaert 1998; Mahrt and Vickers 2004)**. More recently, **Jiménez et al. (2012)** corrected these schemes to make them more universally applicable based on atmospheric conditions.

The parameterizations of the calculation of the height of the boundary layer and surface fluxes are interconnected, like most physical schemes in a model. As they contain many hypotheses and parameters, choosing to apply them requires a knowledge of published work in order to optimally combine the parameterizations.

4

Emissions

4.1. Introduction

Emissions represent the single source of pollutants in the atmosphere. They are variable in time and space depending on their origin *(Granier et al. 2019)*. Emissions are almost impossible to measure: we measure a concentration in the air that has already undergone chemical reactions, transport and deposition. They must therefore be estimated from other information. In this chapter, we will list all emissions of atmospheric pollutants and see the various uncertainties attached to these emissions depending on the method used to estimate them.

Emissions can be separated into two main families: **anthropogenic** emissions and **natural** emissions. Anthropogenic emissions represent human activities, whether related to industry (heavy industries, steel, etc.), transport (roadways, maritime, aviation, rail, etc.), agriculture (livestock and fertilizers) or the residential sector (heating). Natural emissions (biogenic, erosion and pollen) are emissions from vegetation (forests and pollen), soil (erosion), oceans (sea salts) and sporadic sources (fire and volcanoes). The various kinds of emissions release almost all the chemical species that are of interest for pollution (apart from mineral aerosols, which are largely created by wind erosion and pollen). There are three types of processes related to these emissions: first of all there is **combustion** with CO_2, CO, COVNMs (benzene, formaldehyde, etc.), hydrocarbons, dioxins, soot particles and sulfur dioxide (SO_2). Then there is **volatilization** with volatile organic compounds (petrol, paints, etc.), fertilizers and the re-emissions of depositions (persistent organic pollutants (POPs)). Finally, there are **mechanical processes** with metals (industrial activities, abrasion of brakes, etc.), mineral dust (suspensions related to traffic, agriculture) and pollen.

The calculation of these emissions leads to the quantification of a flux (mass by volume and by time). In order to estimate this flux, we must spatially constrain emissions in the following aspects: horizontally (using the databases from the emission zones), vertically (depending on the type of source), temporally, chemically

Modeling of Regional Atmospheric Pollution,
Laurent MENUT. © ISTE Ltd 2024.

by species and by mass (depending on the quantity emitted). For each of these dimensions, the calculation can be constrained by using in situ measurement data or satellite data. We will then frame hypotheses regarding the emission/concentration transition, as it is impossible to directly constrain the emitted fluxes before they undergo atmospheric chemical reactions. Figure 4.1 summarizes the two methodological concepts for determining emissions:

– The *bottom-up* approach is the processing of information that represents the reality of an emission. Based on economic or geophysical data, an emission model aggregates all available information to reconstitute the desired flux. This flux is then used by a chemical transport model, which will calculate atmospheric concentrations. The relevance of the calculation chain will be evaluated by comparisons between the model and the observations.

– The *top-down* approach can take two forms: (i) directly using concentrations to evaluate emissions (this is generally done using satellite data), or (ii) through inverse modeling by comparing simulated concentrations and observed concentrations to modify the emitted fluxes (see section 8.2).

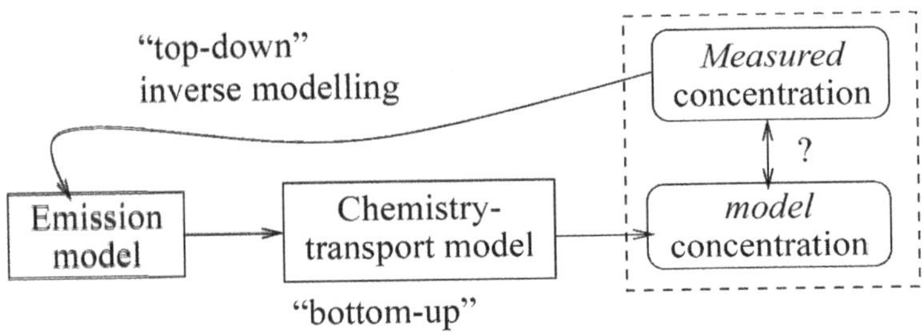

Figure 4.1. *Principle underlying the calculation of emissions using bottom-up or top-down approaches*

Emissions must be calculated with a model (EM) even before the chemical transport model (CTM) is established. Between the data used and the EM model, we have large uncertainty, from the start, regarding the fluxes that the CTM will use. The only evaluation possible is to compare the observed concentrations and the modeled concentrations: we therefore multiply the uncertainties between the EM and the CTM. Finally, we adjust the EM so that the CTMs provide the best possible comparisons between the observed and simulated concentrations. The emissions used cannot be directly transferred from one CTM to another as they depend directly on the CTM that was used to calibrate the emissions.

4.2. Anthropogenic emissions

| Open-access publications: *Im et al. (2018b).*

It is essential to quantify anthropogenic emissions in the study of atmospheric pollution, as these are the only source of pollution caused by humans that can therefore be potentially reduced. Logically, the source of pollution is localized to sites of human activity and therefore the sources are concentrated in large urbanized and industrial centers. Among all the emissions to be calculated to model pollution, these anthropogenic emissions are the least "physical". They are not very dependent on weather conditions or soil type. They are primarily fluxes estimated based on an inventory of human activities.

4.2.1. *Emission inventories*

All modeling systems use databases prepared separately. These databases are difficult to establish as they are constructed using emission inventories organized by species and activity sectors *(Klimont et al. 2017)*. These approaches are the "bottom-up" approaches that compile information on the origin and nature to finally establish an "inventory" of emissions. This laborious inventorying task explains why these bases are relatively static and not regularly updated. Historically, emission databases were regional or continental. This type of database continues to be updated as it represents the dimension that is of interest to forecasting organizations, such as the accredited air quality monitoring associations (AASQA) or the *European Monitoring and Evaluation Programme* (EMEP) *(Tørseth et al. 2012)* used in western Europe. Today, however, developments converge toward unique and global inventories on a kilometric scale. This makes it possible to ensure consistency in the masses emitted between countries and between continents. These unified inventories are developed and maintained by international work groups, such as the EDGAR database[1] and the Copernicus Atmosphere Monitoring Service *(Granier et al. 2019)*.

Emission inventories are issued by chemical species (or grouped species and activity sector). The most widely used nomenclature is the European norm, "Selected Nomenclature for Air Pollution" (SNAP), comprising 11 sectors of activity. The codes used are given in Table 4.1. More recently, institutes working on these emissions, such as the EMEP, implemented the GNFR nomenclature, modifying the list of sectors. This made it possible to go from 11 to 18 sectors, fine-tuning the information. The following inventories are usually used. The EMEP inventory[2] has a resolution of $0.1° \times 0.1°$, and an inventory for each year from 1990 to 2017. Although it only

1 Available at: https://edgar.jrc.ec.europa.eu/.

2 Available at: https://www.ceip.at/.

covers Europe, it is widely used. It lists the following species: CO, NH_3, non-methanic volatil organic compounds (NMVOC), NO_x, SO_x, $PM_{2.5}$ and PM_{10}. The masses given are annual. The CAMS inventory[3] has the same resolution but gives monthly masses. It is also only available for the years after 2000. The inventory covers the same species, as well as black carbon (BC) and organic carbon (OC), and it is a global inventory. The HTAP global inventory[4] from the EDGAR database only has an inventory for the year 2010. It also has a resolution of $0.1° \times 0.1°$. These inventories are generally used separately, but certain studies carry out comparisons between them, such as **(Elguindi et al. 2020)**: although they are supposed to represent the same thing, there is a large variability between inventories, and the differences range from single to double.

#	Activity sector
01	Combustion in the production and transformation of energy
02	Non-industrial combustion plants
03	Industrial combustion plants
04	Industrial processes without combustion
05	Extraction and distribution of fossil fuels and geothermal energy
06	Use of solvents and other products
07	Road transport
08	Other mobile sources and machinery
09	Waste treatment and disposal
10	Agriculture
11	Other sources and sinks (nature)

Table 4.1. *Sectors of activity in the European SNAP norm*

The inventories are not updated every year. However, in relation with policies to reduce pollution, these emissions vary from one year to another. We must therefore apply corrective factors in order to conserve annual trends. Across Europe, for example, these trends are quantified in *Vestreng et al. (2009)*; in China, these trends are quantified in *Zheng et al. (2018)*. Studies also use multi-annual time-series of satellite measurements to estimate emission trends. Using the hypothesis that surface concentrations of nitrogen oxides, as well their emissions, change close to their sources, *Konovalov et al. (2008)* has quantified the inter-annual variability in EMEP emissions in Europe. The study carried out by *Hoesly et al. (2018)* also makes it possible to quantify long-term trends in emissions, from 1750 to 2014, using a reconstruction of the evolution of industry.

3 Available at: https://eccad.sedoo.fr/.

4 Available at: https://edgar.jrc.ec.europa.eu.

There are also emission inventories for future scenarios, like climate change modeling exercises. Let us first make note of the "representative concentration pathways" (RCP) scenario *(Moss et al. 2010)*. These scenarios offer data on emissions and changes in landcover depending on the level of radiative forcing (in $W.m^{-2}$) expected for 2100. The computed forcing is based on greenhouse gas pollutants and does not take into account any eventual changes in land use. Various scenarios are suggested: RCP2.6, 4.5, 6.0 and 8.5. More recently, for the CMIP6 exercise, the "shared socio-economic pathways" (SSP)[5] scenarios were put forth for a deeper understanding of the feedback between climate change and socioeconomic factors by constructing scenarios that ensure coherence between both these factors. Many studies have worked toward establishing the most realistic scenarios possible, even if the future is highly uncertain. Examples of such studies include *Gidden et al. (2019)* and *Feng et al. (2020)*.

4.2.2. *Emission cadastres*

In order to use these data in a chemistry-transport model, these inventories must be converted into a format compatible with the model – *emission cadastres*. The constraints here are that we must have emission data for each time step of the forcing in the model (generally hourly), for each grid cell, and formulated with model species. We must therefore carry out different calculations to shift from the inventory to the cadastre: spatialization, temporalization and chemical speciation, as shown in Figure 4.2.

There are two major uncertainties in the calculation of these emissions: (i) the inventory data itself, constructed bottom-up from lists and statistics, and (ii) the errors made during the spatiotemporal projection of the inventory toward the cadastre required for a simulation. There are models that can carry out these calculations. For example, the Hermes code *(Guevara et al. 2020)*, the EmiSurf code **(Menut et al. 2012)** and the DPEC code for China *(Tong et al. 2020)*.

4.2.2.1. *Spatialization*

Horizontal

The emitted masses may be available depending on the very varied surface areas of the grid cells in the model that is being used. This may be a Cartesian mesh with an offset, with different resolution, or masses based on regions or the area of communes. Thus, these masses must be projected from one (any) grid to another (the CTM grid), while conserving mass. In the case of a projection onto a mesh with better resolution, the calculation may use "subgrid" or proxy information: demographic data, the type of surface to accurately represent roads for traffic emissions, etc. This aspect is especially

5 Available at: http://www.ipcc.ch.

important since diluting emissions in a grid cell may modify the chemical regime. An example is given in Figure 4.3, as used in the EmiSurf model developed to prepare emissions from the CHIMERE model *(Mailler et al. 2017)*.

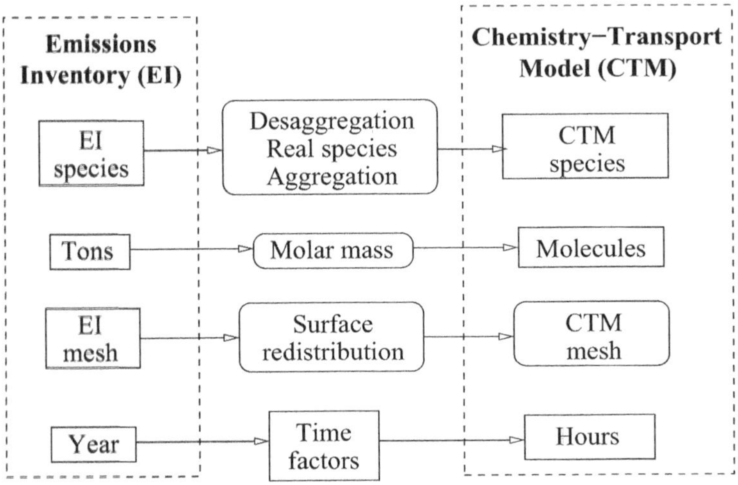

Figure 4.2. *Principle for the calculations to move from an inventory to a cadastre of anthropogenic emissions*

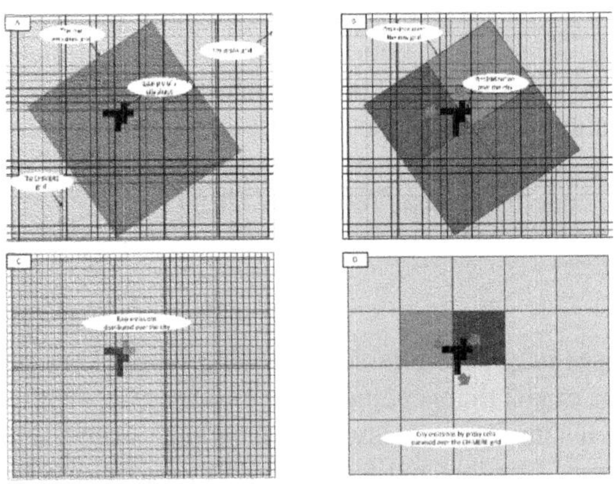

Figure 4.3. *Principle of projecting anthropogenic emission masses from the inventory mesh (grid of origin) onto the cadaster for the chemistry-transport model. Adopted from* Mailler et al. (2017). *For a color version of this figure, see: http://www.iste.co.uk/menut/modeling.zip*

For industry or residential activity sectors, the spatialization information is quite easily accessible as it consists of listing occupancy maps, bearing in mind that these activities are not very displaced over time. The spatialization of vehicular traffic is trickier, because although roads and freeways are well identified, the quantity of the emitting vehicles is very variable and hard to predict.

Inventories usually provide data in three forms: (i) **surface sources**, representing industrial, urban and agricultural zones; (ii) **line sources**, representing waterways, freeways, streets and railway lines; and (iii) **point-like sources**, representing highly localized industrial sites, such as factories. Each type of source will be spatially re-projected onto the mesh used for the model, ensuring that its representativity is optimally retained, as is its mass. The difficulties that arise may be, for example, (i) the surface source has a part that falls outside the target model mesh and (ii) the line sources (like boats) yield an emitted mass over the entire trajectory of the boat, including outside the target domain.

Vertical

The masses are given in two dimensions and must be vertically redistributed. Figure 4.4 presents the vertical profile applied across seven levels, from 20 to 1100 m, as per the EMEP nomenclature. For instance, it can be seen that SNAP1 (industrial combustion) is mainly emitted by chimneys and the highest redistribution percentage is located between 300 m and 600 m. On the contrary, for SNAP7 (vehicular traffic) 100% of the emitted mass remains at the first level.

4.2.2.2. *Temporalization*

Raw data are generally available in the form of mass per year. We must therefore "redistribute" it to obtain hourly fluxes (the general case of CTMs used for regional pollution). This redistribution follows several successive steps: going from mass per year to mass per month, to account for seasonality; then for each month looking at the type of day (to account for variability over the week); and then estimating a diurnal profile (for daily variability). This temporal redistribution may also depend on the site: two different cities will not have the same daily activity in terms of traffic. Figure 4.5 offers an example of daily profiles that make it possible to go from mass per day to an hourly diurnal profile, depending on the day of the week, for London and Madrid. The diurnal cycle shows high variability between a weekday and Saturday and Sunday.

The methodology used to estimate these profiles depends on the activity sector. For an industrial source, it is possible to gain information on the activity of the factory and thus obtain a diurnal profile. For vehicular traffic, depending on the number of vehicles, the concentration measures closest to the sources were used as proxies. "Urban" NO_2 measurements are used in **Menut et al. (2012)** to differentiate between cities. The study by *Crippa et al. (2020)* compiles many other studies to aggregate a database with temporal profiles on a global scale. The authors specify that they have created a database for the year 2005, considered as an average and representative year at the meteorological level. Indeed, using the measurement of concentrations in air as proxies includes the fact that these measurements have a variability that is linked to the emission and also, potentially, linked to the dynamics. The study by *Tunved and Ström (2019)* shows that there may also be seasonal variability in the size distribution of aerosols, depending on the emission trends from year to year: this kind of variability currently corresponds to a fine-tuning that is rarely considered in emission models.

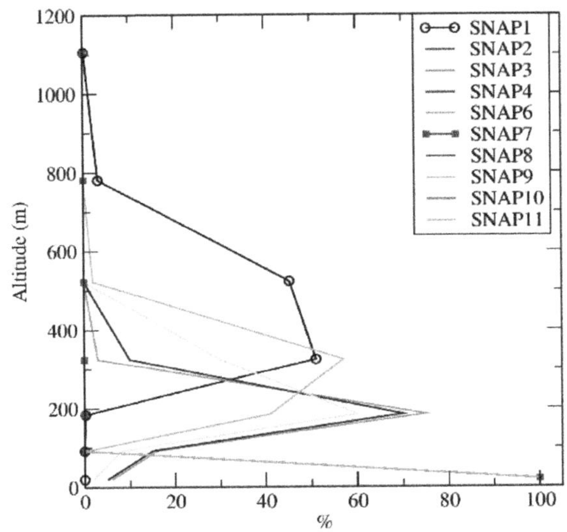

Figure 4.4. *EMEP profiles for the vertical redistribution of surface data. As per data extracted from EMEP data. For a color version of this figure, see: http://www.iste.co.uk/menut/modeling.zip*

4.2.2.3. *Chemical speciation*

Emission mass data are obtained for species that represent real species (such as SO_2) or groups of real species (such as NO_x and VOCs). Further, these masses are given by activity sector, while the model uses only a single chemical species. We must therefore construct de-aggregation matrices and then matrices for the re-aggregation of real species to model species. A two-step process must be applied: first, de-aggregating the "model" species from the inventory into "real" species; next, aggregating these "real species" into model species for the CTM chemical scheme used. Depending on the model, there are two types of re-aggregation carried out: a molar mass weighting factor or a reactivity weighting factor. For a reactivity weighting factor, α_i, we use the following relation:

$$\alpha_i = \frac{1 - exp(-k_i \times I_{OH})}{1 - exp(-k_{Model} \times I_{OH})} \qquad [4.1]$$

$$E_{Model} = \sum_i \alpha_i E_i \qquad [4.2]$$

where k_i is the OH reactivity rate of the real compound and k_{model} is the OH reactivity rate of the real species. I_{OH} is the integration over one day of the OH concentrations in ppt.mn^{-1}. It can be noted that most inventories provide masses of NO_x and SO_x. In CTMs, the usual decomposition of NO_x is 95% NO, 4.5% NO_2 and 0.5% HONO, and the decomposition of SO_x is 99% SO_2 and 1% H_2SO_4. These properties may be modulated depending on the sector of activity.

It must be noted that these chemical conversion steps must be carried out at the end, or in any case, after temporalization, since most temporal profiles of the factors are given by activity sector. We must therefore apply these before losing this information during the shift toward the model chemical species.

Anthropogenic emissions remain the principal unknown today, when modeling regional pollution. The finer the resolution of the model, the more impact this uncertainty has on the results: we must describe roads, freeways, streets, buildings, etc., as fully as possible. Further, increasing the temporal resolution requires a knowledge of traffic up to present-day constraints where it is impossible to predict bottlenecks from one day to the next. For operational forecasting, the large temporal variability in the sources is going to remain an obstacle for some time to come. It seems hard to envisage assimilating real-time traffic data over large regions using the modeling systems that are currently in use.

4.3. Biogenic emissions

Open-access publications: *Guenther et al. (2006).*

Biogenic emissions represent gaseous emissions, principally VOCs (isoprene, terpenes, pinene), produced by forests *(Simpson et al. 1999)*. In order to estimate the flux of these emissions, we must have information on the vegetation and meteorology. The emitted fluxes depend directly on the chemical species emitted, on the plant or tree, on the temperature and on solar irradiation. For this last parameter, calculation schemes used for biogenic emissions have defined **photosynthetically active radiation** (PAR). Thus, unlike anthropogenic emissions, we have emissions that will depend both on the type of surface and on meteorology.

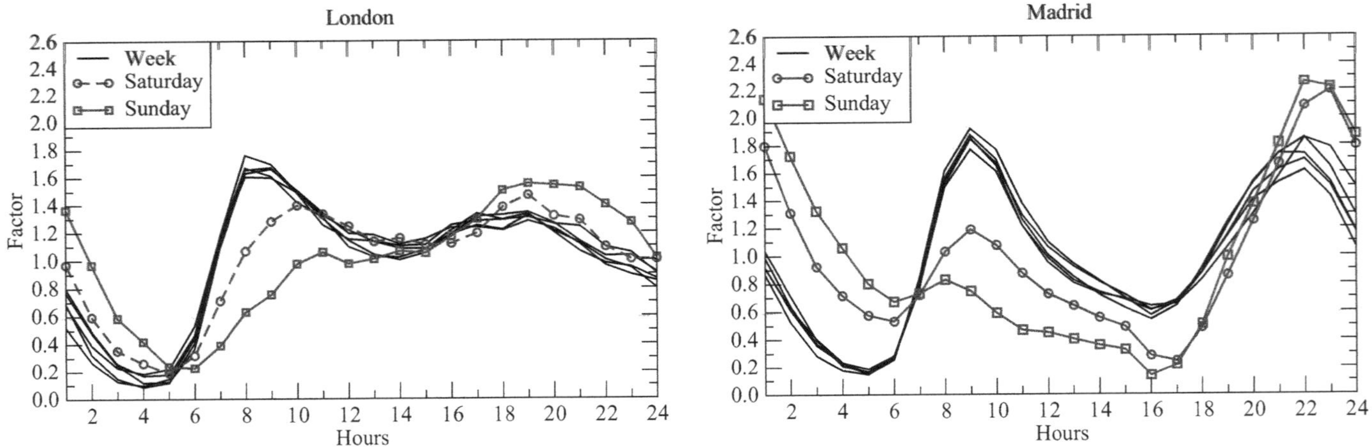

Figure 4.5. *Coefficient representing the daily cycle of emissions linked to traffic for the cities of London and Madrid. For each city, the diurnal profiles have been obtained with measurements from "traffic stations", with an average over several months for each Monday, Tuesday, etc. Data taken from the study by* **(Menut et al. 2012)**. *For a color version of this figure, see: http://www.iste.co.uk/menut/modeling.zip*

4.3.1. *Flux parameterization*

The general process is the chemical formation of VOCs in leaves, and these are then diffused into the air from the surface of the leaf. There are not many different models that model this process. Most CTMs use MEGAN (*Model of Emissions of Gases and Aerosols from Nature (Guenther et al. 2006)*). This model calculates the fluxes emitted from surface data (supplied with the model) and the meteorological parameters that must be calculated during the chemistry-transport simulation. We must also mention the BEIS3 model, today integrated into the emission calculation system *Sparse Matrix Operational Kernel Emissions* (SMOKE), used for the CMAQ model **(Pierce et al. 1998)**. The emitted flux is then estimated according to the relation:

$$ER_i = \epsilon i \times \gamma_i(T, q, PPFD, LAI) \times \rho_i \qquad [4.3]$$

with ER_i ($\mu g.m^{-2}.h^{-1}$) being the emission rate of the chemical species, i, ϵ_i ($\mu g.m^{-2}.h^{-1}$) being an emission factor for the standard meteorological conditions of the canopy, γ_i (without unity) being a factor of the variability around these standard conditions, depending on the type of plant the meteorological conditions, and ρ_i being a factor that takes into account the production/loss balance of the chemical species in the canopy. The concept of a "standard canopy" corresponds to the following conditions: a temperature of 303 K, a photosynthetic flux density (PPFD) of 1,500 $\mu mol.m^{-2}.s^{-1}$ at the peak of the canopy, a foliar index *leaf area index* (LAI) of 5 $m^2.m^{-2}$ and a canopy that is 80% mature leaves, 10% new leaves and 10% old foliage. In the canopy, the factor ρ_i is considered to be equal to 1. The entire parameterization consists of estimating the factors in variability around the mean state.

The general principle underlying this calculation is presented in Figure 4.6. Using this calculation principle, MEGAN makes it possible to calculate the emitted fluxes for around 20 biogenic VOCs. The calculated flux is used at the first vertical level of the models: the injection profile into the atmosphere, used for anthropogenic emissions or biomass fires, are not taken into consideration. However, this may be refined by more realistically taking into consideration the height of the canopy.

4.3.2. *Databases*

In order to calculate the term ϵ_i in standard meteorological conditions, we must have databases with information on the surface, with information spatialized based on the plant type. These bases are global and are based on the interpretation of satellite data. In the case of MEGAN, the databases are supplied with the model. There are also in situ measurements that enable us to link satellite measurements (which give spatial variability) with a measurement value. For example, to estimate the emission factor for isoprene, *Guenther et al. (2006)* collected results from the field study measurement campaign, as shown in Figure 4.7. It can be noted that even if this represents data from many different campaigns, these constraints remain rare across the globe.

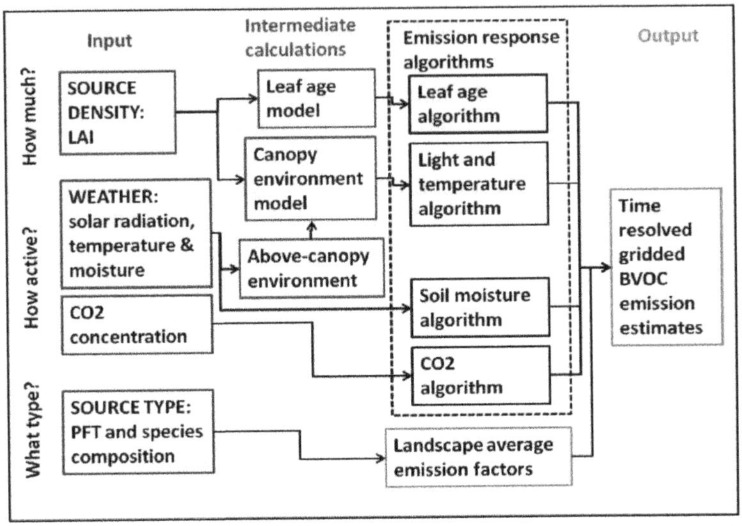

Figure 4.6. *Principle underlying the calculation of biogenic emissions. Figure adopted from* Guenther et al. (2012). *For a color version of this figure, see: http://www.iste.co.uk/menut/modeling.zip*

Figure 4.7. *Measurements used to estimate the emission factor for isoprene. Figure adopted from* Guenther et al. (2006). *The colors correspond to a classification into eco-regions that was used in the article. For a color version of this figure, see: http://www.iste.co.uk/menut/modeling.zip*

To reduce the amount of constraining data, the plants are categorized. We define *plant functional types* (PFT) by creating categories of vegetation that have the same type of seasonal cycle and the same kind of foliage or needles, and therefore emit similar quantities of VOCs. An example of the emission factor for isoprene is given in Figure 4.8. These factors are estimated with a monthly frequency in order to obtain the seasonal factor on the evolution of leaves in trees and therefore their productivity.

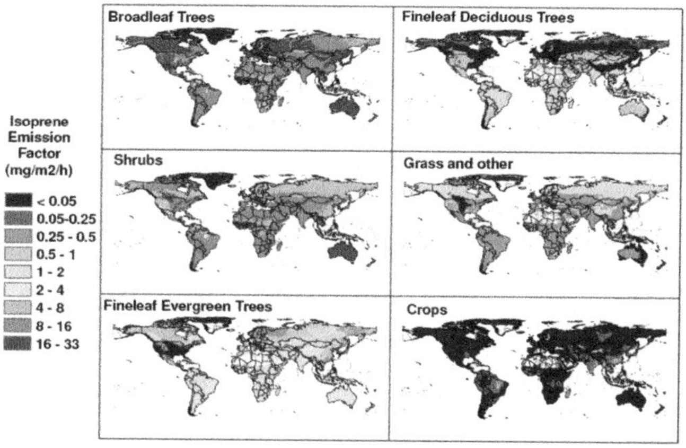

Figure 4.8. *Global isoprene emissions: emission factors by type of emitting plant and emission fluxes deduced based on the season. Figure adopted from* Guenther et al. (2006). *For a color version of this figure, see: http://www.iste.co.uk/menut/modeling.zip*

4.3.3. *Intercomparison and sensitivity*

Various studies intercompare the calculation of biogenic emissions and quantify the sensitivity of the calculations of the flux concentrations. *Emmerson et al. (2018)* compared two models in Australia: MEGAN and the Australian ABCGEM. Their sensitivity study showed a high interdependence of the results on the emission factors of monoterpene, especially its dependence on sunlight. *Jiang et al. (2019)* compared MEGAN and PSI in Europe and demonstrated that factors related to vegetation lead to a difference of a factor of 3 in the isoprene and monoterpene calculations. This has an impact of the order of 10 ppb on ozone and also a non-negligible impact on the formation of secondary organic aerosols. Many studies use the biogenic emission calculation model to study long-term trends of these emissions. *Sindelarova et al. (2014)* calculated 30 years of flux over the globe and showed its seasonal variability as well as inter-annual variability, with a marked decennial cycle. This variability is only due to meteorological variability. From the point of view of global uncertainty, they compiled many studies of the same type and showed that the estimation of the emitted mass of isoprene ranged from 400 to 700 Tg per year.

Biogenic emissions are responsible for half the atmospheric emissions. The calculations used to model them are a mixture of information on the surface and on meteorology. Satellites make it possible to estimate surface data: the databases advance regularly in spatial and temporal resolution, and in the quality of the estimated values. The calculation of the emitted fluxes is itself directly dependent on meteorology and therefore must be directly included in the chemistry-transport model.

4.4. Emissions from biomass burning

Open-access publications: *Reid et al. (2005a)*.

Emissions from fires are among the hardest to estimate as they involve several problems: (i) they can depend on anthropogenic activities (start of the fire, type of surface, type of vegetation) or natural causes (lightning), as well as on meteorology (propagation of the fire); (ii) they are difficult to measure (impossible on the ground, difficult by satellite); and (iii) they are highly variable across time and space, so the concept of an inventory is impossible. The calculation of emissions from fire is therefore one of the most parameterized and therefore one of the most uncertain.

These emissions have been intensely researched and constitute one of the sources that can be calculated with a large number of different models. There are articles that give details on these processes, such as *(Reid et al. 2005a)* on the physical properties of particles from fire, and *(Reid et al. 2005b)* on their optical properties. These emissions have a significant impact on the composition of the atmosphere. Unlike other kinds of emissions, these are very intense but highly localized. Finally, by comparing emissions from fire with anthropogenic or biogenic emissions, their mean impact is smaller *(Rea et al. 2015)*. Aerosols also have an impact on the radiation balance: the study by *Brown et al. (2021)* showed that aerosols from fires were not well represented for their radiative impact and that most models considered them to be more absorbent than they are in reality, which affected the results from the climate change modeling exercise.

Biomass combustion can be divided into two categories: (i) complete combustion leads to the formation of CO_2 and H_2O; (ii) incomplete combustion will produce many trace gases and aerosols, depending on the type of fuel, the quality of the combustion and the composition of ambient air in some cases (e.g. reciprocating engine). In the case of incomplete combustion, which is the case with most biomass fires, the compounds will be emitted based on the ambient air and the elements present in the fuel. Biomass combustion will generate carbon dioxide (CO_2) and carbon monoxide (CO), volatile organic compounds (benzene, formaldehyde, etc.), semi-volatile organic compounds and non-volatile organic compounds (polycyclic aromatic hydrocarbons, dioxins, etc.), soot particles, sulfur oxides (SO_2 and SO_3) and halogen mineral compounds (HCl) and mercury compounds (HgO, $HgCl_2$ etc.). After emission, the composition of the plume will also be modified, with active chemistry being present. This may have an impact on the concentrations and optical properties of aerosols, and therefore the optical density *(Konovalov et al. 2017)*.

4.4.1. *Surface emission processes*

The modeling of this kind of fire has similarities with biogenic emissions: the basis of any calculation is the high-resolution mapping of soil and vegetation around the globe: we must also have essential information on the tree, this time to propose emission factors during its combustion. However, we also need additional information: a vegetation model to know the quantity of biomass and satellite data to know the location of active fires. The modeling of emission fluxes from biomass burning takes the following form as per *(Heinold et al. 2011)*:

$$M_x = \sum_k [M_{biomass,k} \times A_{burn,k} \times EF_{x,k} \times BE_k] \quad [4.4]$$

where x is the chemical species emitted and k is the biome (tropical forest, desert, etc.). M_x is the flux, expressed in mol day^{-1} m^{-2} for gases, and in g day^{-1} m^{-2} for aerosols. $A_{burn,k}$ is the burnt surface, BE_k is the burning efficiency and $EF_{x,k}$ is an emission factor. $M_{biomass,k}$ is the available fuel load. There is also an approach based on satellite measurements of the fire radiative power (FRP), which makes it possible to parameterize the quantity of fuel consumed *(Kaiser et al. 2012)*. With this information, the flux can also be modulated by the satellite products of "active fires", which yield a thermal anomaly, and therefore indicate the presence of a fire with high temporal resolution. For many current models, the burnt surfaces and active fires, $A_{burn,k}$, are estimated a posteriori, using data from a MODIS satellite *(Giglio et al. 2006)*. Not having enough information on the localization a posteriori demonstrates the difficulty in forecasting these emissions.

There are many models that can calculate these surface fluxes. Many studies intercompare this range of models and quantify their differences: depending on the hypotheses used for calculations, and based on the parameters on vegetation, the burning efficiency or other facts, comparisons can range from single to double differences. *Turquety et al. (2014)*, for instance, compared the GFEDv3, FINNv1, GFASv1 and APIFLAMEv1 models for the Euro-Mediterranean region. *Pan et al. (2020)* compared six databases of emissions from fires at a global scale and showed that there is a factor of 4 between the different datasets. These datasets have highly variable resolutions from 0.5° to 1 km. It must be noted that among these models, the GFAS model by ECMWF *(Kaiser et al. 2012)* assimilates data in the chemistry-transport model and thus offers a more reliable product as it is corrected by independent information. *Pereira et al. (2016)* compared inventories more regionally, across Brazil, with 3BEM, GFAS and FINN, and also concluded that while the spatial distribution of sources is consistent between datasets (impact of the satellite forcing), the emitted masses are very different (hypotheses on the fluxes).

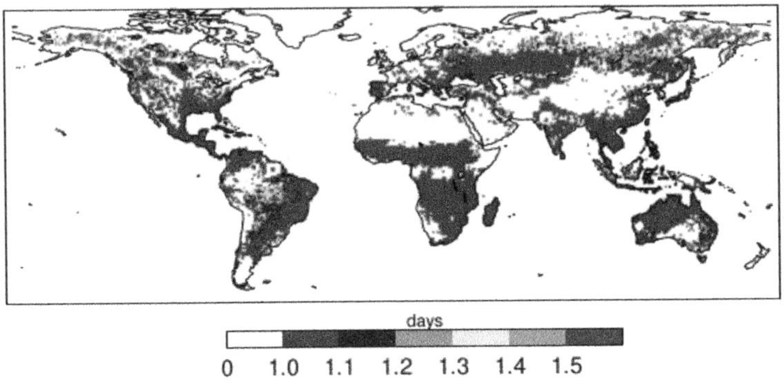

Figure 4.9. *The mean fire persistence, in days, for the period 2001–2004, analyzed from the active fire product from Terra MODIS. Figure taken from* Giglio et al. (2006). *For a color version of this figure, see: http://www.iste.co.uk/menut/modeling.zip*

4.4.2. *Pyroconvection*

If it is difficult to estimate fluxes emitted on the surface, it is all the more difficult to estimate the injection of these fluxes into the atmosphere. Unlike other emissions, anthropogenic or biogenic, the surface flux is not released from the surface at ambient temperature – the product of the fire is emitted into an atmosphere where the temperature can reach a few hundred degrees. The meteorology is therefore completely disrupted and pyroconvection takes place. The biomass fire product may be rapidly injected a few thousand meters from the surface, above the boundary layer. We must therefore develop specific parameterizations to estimate the vertical injection profile. This profile is estimated through two distinct steps: (i) the injection height H_p and (ii) the shape of the injection profile.

The **injection height** is parameterized and there are many schemes for this *(Paugam et al. 2016)*. These schemes can then be directly compared with the measurements like with instruments on the CALIOP satellite or the *Multi-angle Imaging SpectroRadiometer* (MISR). One of the most widely used schemes is that given by *Sofiev et al. (2012)*. The H_p calculation is based on the estimation of the convective available potential energy (CAPE):

$$H_p = \alpha H_{abl} + \beta \left(\frac{P_f}{P_{f0}} \right)^{\gamma} exp\left(-\frac{\delta N_{FT}^2}{N_0^2} \right) \quad [4.5]$$

where $\alpha = 0.24$, $\beta = 170\,\text{m}$, $\gamma = 0.35$, $\delta = 0.6$, $P_{f0} = 10^6\text{W}$ and $N_0^2 = 2.4 \times 10^{-4}\ \text{s}^{-2}$. The FRP, P_f, is expressed in W (with 1 W = 1 J s^{-1} = 1 m^2 kg s^{-3}). N_{FT} is the Brünt–Vaisala frequency in the free troposphere. An empirical correction has been brought into account for the underestimation of the FRP by MODIS in the case of "large fires" *(Veira et al. 2015)*:

$$P_f^* = P_f \times \left(\frac{H_p}{H_{deep}} \right)^{\epsilon} \quad [4.6]$$

with $\epsilon = 0.5$ and $H_{deep} = 1{,}500\,\text{m}$.

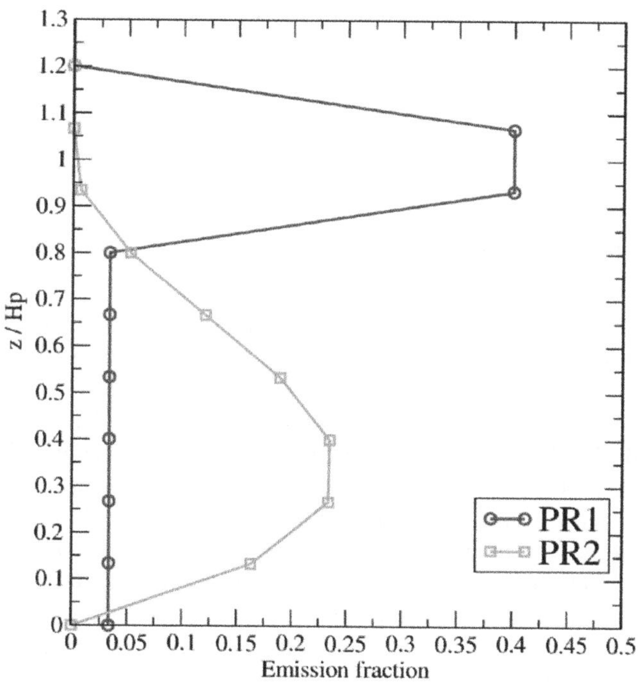

Figure 4.10. *Vertical profiles for the injection of biomass fire products in the troposphere. Figure taken from* Menut et al. (2018). *For a color version of this figure, see: http://www.iste.co.uk/menut/modeling.zip*

The calculation of the **shape of the profile** is also parameterized. Just as with the injection height, this parameter is highly sensitive: depending on the altitude at which a pollutant is injected, it may be transported in any direction, and its deposition may be more or less delayed. Most of the models assume a homogeneous injection into the boundary layer, or at a height H_p from the surface. *Sofiev et al. (2013b)* distribute the flux homogeneously from $H_p/3$ to H_p. Other models use more complex approaches, like thermaps, which are adapted to the problem posed by pyroconvection *(Freitas et al. 2007; Rio et al. 2010)*: an approach with a heavy calculation time and which is ultimately not used very much. Finally, other approaches consider K_z-shaped profiles with a maximum at the altitude $H_p/2$ *(Raffuse et al. 2012; Veira et al. 2015)*. Both these shapes are represented in Figure 4.10 and intercompared in *Menut et al. (2018)*.

Emissions from fires are complex to estimate as they depend on many non-measurable factors: When will a fire start? What are the species emitted? In what quantity? While on an annual average these emissions are not the principal source, they are violent and intense wherever they are localized. Short-term forecasting of these emissions is still a challenge today, as it is not possible, on a regional scale, to accurately calculate the propagation of a fire or its duration. Certain parameters are still not taken into account today in models, such as the consideration of the burnt area from one year to the next, the impact of a (small scale) fire on the meteorology on a synoptic scale, or again the precise radiative balance of a plume of products of the fire at altitude.

4.5. Mineral dust emissions

Open-access publications: *Schepanski (2018)*.

Mineral dust emissions represent a large part of the mass of aerosols emitted across the globe **(Ravi et al. 2011)**. This mass, however, remains an uncertain quantity, with the order of magnitude of the difference varying between models *(Shao et al. 2011)*. These aerosols have a potentially significant health impact on the respiratory system **(Morman and Plumlee 2013)**. They also have a high impact on the radiative balance **(Carslaw et al. 2010)**, both in the short term, during the transport of dense plumes, and in the long term from a climatological point of view. We differentiate between "natural" emissions (as in the Sahara and Sahel, in Africa) and "anthropogenic" emissions (from industrial sites or roadways) **(Engelstaedter et al. 2006)**. The largest emitted mass is "natural", but "anthropogenic" emissions are closer to inhabited areas and therefore must also be considered in the context of atmospheric pollution monitoring. Just like the emissions arising from fires, these emissions have been the subject of many research studies and there are many models. Available articles include **Ravi et al. (2011)**, **Shao et al. (2011)**, **Choobari et al. (2014)** and **Middleton (2017)**.

Mineral dust emissions are observed when the soil is arid and the wind near the surface is strong enough to suspend the particles in the surface layer. This emission process, called wind erosion, will mainly depend on the type of surface and meteorology. Observations and simulations in wind tunnels have shown that the emission can be broken down into two phases: a horizontal flux (saltation) and a vertical flux (sandblasting). It is tricky to model this process as it covers multiple spatial scales: close to the surface the erosion process has a centimetric threshold. The injection of the particles into the surface layer and then the vertical mixing in the boundary layer are on a larger scale of the order of the surface heterogeneity. There are many parameterizations for the calculation of these emission fluxes, with approaches that can be very varied. However, the common point in all these approaches is that they depend on thresholds. The concept of a threshold itself is highly dependent on the horizontal resolution when the process is modeled. Therefore, the entire difficulty in calculating these emissions is linked to the fact that we are working on **a sensitive process, with a threshold, highly dependent on the quality of meteorological forcings and the model's resolution**. Three ingredients are required to parameterize theses emissions: (1) the surface and its roughness (presence of vegetation, rocks, etc.); (2) the composition of the soil (minerology); and (3) the meteorology close to the surface with the turbulence, soil moisture and wind speed.

4.5.1. *Characterization of the surface, the soil and roughness*

The (*landuse*) and soil type will govern whether emissions are possible or not. The models use regional or global emissions, established from satellite data. These are the same databases that are used for the meteorological models or CTMs. Among the databases used, for landuse we have the following: (1) U.S. Geological Survey Landuse (USGS) (**Loveland et al. 2000**), with a resolution of 1 km across the globe, and 28 surface classes (urban, grass, copses, lakes, etc.). (2) GlobCover Land Cover: a map with 300 m resolution (**Bicheron et al. 2011**). It comprises 22 surface classes and was designed based on information from the MERIS sensor on the ENVISAT platform. Figure 4.11 presents percentages for surface type in two categories: the urban area and barren soils in the USGS database. The mesh resolution here is 60 km.

Soil type must be characterized to find out whether the soil is made up more of sand, silt or clay. Mineral aerosols suspended in air are principally composed of sandy soils and clayey soils (after fragmentation) (**Zobler 1986**). For the Sahara, **Chatenet et al. (1996)** proposed a classification into four elements, depending on the median diameter of soils, such that: $D_p = 125\ \mu m$ for aluminosilicates (composed of clay primarily), $D_p = 210\ \mu m$ for fine sand (composed of quartz and other minerals) and $D_p = 690\ \mu m$ for coarse sand (quartz). More recently, models have tried to refine the mineralogical composition of soils, like those proposed by **Claquin et al. (1999)**, *Balkanski et al. (2007)* and *Menut et al. (2020b)*. This makes it possible to access the iron concentration (for biogeochemical modeling), better account for the optical properties of the minerals, have transport with particles of different densities, and therefore model deposition better. Further, this enables an additional comparison with observations of atmospheric concentrations or deposits, which may be focused on iron or calcium. The soils in the calculations can be characterized using global databases like the USGS database STATSGO-FAO with a 1 km resolution. This database contains 19 different soil types, which are combinations of sand, silt and clay.

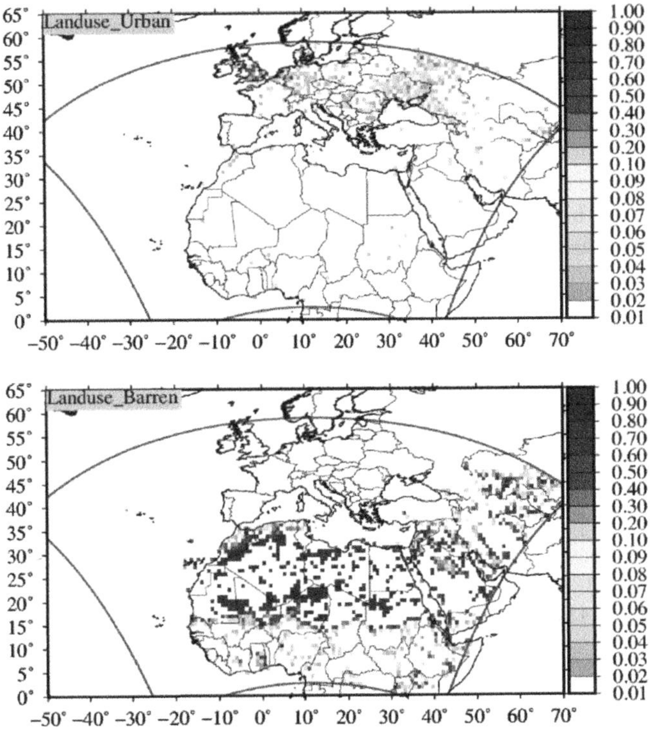

Figure 4.11. *Occupancy maps for surfaces and fractions (between 0 and 1) of urban landuse or barren soil, created using USGS data over a domain containing northern Africa and Europe. For a color version of this figure, see: http://www.iste.co.uk/menut/modeling.zip*

Calculating emissions also requires characterizing the saltation roughness length z_0. This quantity characterizes the roughness over a small spatial scale (centimetric) and is thus different from "dynamical" roughness, as defined in meteorological models. The databases used to define this parameter are constructed from parameterizations interpreting satellite data. For example, the GARLAP (Global Aeolian Roughness Lengths from ASCAT and PARASOL) database developed by

Prigent et al. (2012). Figure 4.12 presents an example in the same simulation domain. The smallest values correspond to the least rough soils, which are therefore likeliest to produce emissions. The modeling may also use erodibility maps. These high spatial resolution maps, which use satellite data, characterize whether or not the surface is erodable, independent of the surface type. This information must therefore be coupled with the soil type to obtain the correct characteristics of this emission **(Beegum et al. 2016)**, **(Menut et al. 2013a)**. Finally, the seasonal nature of these emissions must also be taken into account when moderating the fluxes using the variability of the vegetation.

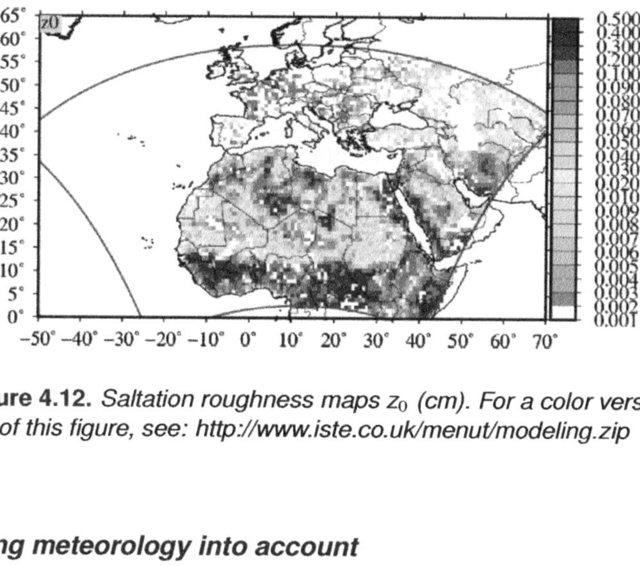

Figure 4.12. *Saltation roughness maps z_0 (cm). For a color version of this figure, see: http://www.iste.co.uk/menut/modeling.zip*

4.5.2. *Taking meteorology into account*

Four meteorological variables are required to calculate mineral emissions: the wind speed at 10 m, U_{10m}; the friction velocity, u_*; precipitation; and soil moisture, q_{soil}. The wind speed is the driving force of emission and choosing to take its value at 10 m from the ground is linked to the parameterizations initially constructed by adjusting the measurements from the networks. This variable is therefore crucial in modeling and its accuracy governs whether or not a good simulation is obtained. Since the wind speed is a highly variable spatially, the larger the grid cell of the model, the less useful the mean wind value in the mesh, as it becomes less and less representative. To counter this problem, models use wind distributions, like the Weibull distribution **(Pryor et al. 2005; Menut 2018)**:

$$p(|U|) = \frac{k}{A}\left(\frac{|U|}{A}\right)^{k-1} exp\left[-\left(\frac{|U|}{A}\right)^{k}\right] \qquad [4.7]$$

with k being the distribution width and A being a scale factor; in our case, the mean wind speed. Precipitation and soil moisture are variables that moderate the flux. Taking precipitation into account makes it possible to consider crustation effects: after rainfall, grains of sand show greater cohesion between themselves than if they have been dry for a long time, and are thus less erodible. To take this effect into account, models like CHIMERE add a function f_p, which is a multiplicative factor to moderate the calculated flux. This factor diminishes with the days that pass without rainfall (**Menut 2018**). The modeling may be simple, for example:

$$f_p = E_{dust} \times \left(1 - exp\left(\frac{-\alpha \times \Delta t_p}{\tau}\right)\right) \quad [4.8]$$

with α being a constant factor such that $\alpha = 6$, Δt_p being the time elapsed since the last precipitation and τ being the period after which the flux is revived at 100% ($\tau =$ 24 h, for example). Soil moisture also modifies emission, as the grains may be stuck to one another. A parameterization used in models to calculate this moderation is the scheme by **Fecan et al. (1999)**. In this case, it is not the final flux that is moderated by the threshold shear velocity, u_*^T, which is increased. To differentiate between dry soil and moist soil, we define threshold shear velocities for each case: u_*^{Td} and u_*^{Tw} (for *dry* and *wet*). We have the relation:

$$u_*^{Tw} = f(w) \times u_*^{Td} \quad [4.9]$$

The factor $f(w)$ is estimated such that:

$$\begin{cases} f(w) = 1 & \text{for } w < w' \\ f(w) = \left[1 + A\,(w - w')^{b'}\right]^{0.5} & \text{for } w > w' \end{cases} \quad [4.10]$$

with A and b' being constants that have to be estimated (**Fecan et al. (1999)** propose $A = 1.21$ and $b' = 0.68$.). w (in kg.kg^{-1}) is the gravimetric soil water content and w' corresponds to the minimum moisture in the soil beyond which the threshold shear velocity increases. The value of w' mainly depends on the clay content in the soil and its value can be estimated so that:

$$w' = 0.0014(\%clay)^2 + 0.17(\%clay) \quad [4.11]$$

One of the parameters in this scheme is the threshold shear velocity. It comes into play when calculating mineral emissions as the parameterizations are expressed in shear velocity, u_* (m/s), diagnosed from the wind velocity at 10 m. Various parameterizations exist for this threshold velocity, denoted by u_*^{Td}. One of the most widely used is the parameterization from **Shao and Lu (2000)**, which is described here. We will have a mineral saltation flux only if $u_* > u_*^T(D_p)$ for a particle with the median diameter D_p. We first define the "smooth" threshold velocity u_*^{Ts} so that:

$$u_*^{Ts}(D_p) = \sqrt{a_n \left(\frac{\rho_p g D_p}{\rho_{air}} + \frac{\gamma}{\rho_{air} D_p} \right)} \qquad [4.12]$$

with the constants $a_n = 0.0123$ and $\gamma = 300$ kg m^{-2}. The average density of the particle is taken to be $\rho_p = 2.65 \times 10^3$ kg m^{-3} and is representative of the majority of emitted minerals, quartz. ρ_{air} is the density of air, here fixed to be $\rho_{air} =$ 1.227 kg m^{-3}. g is acceleration due to gravity. We therefore have u_*^T such that:

$$u_*^{Td}(D_p) = \frac{u_*^{Ts}(D_p)}{f_{\text{eff}}(z_0, z_{0s})} \qquad [4.13]$$

f_{eff} represents the drag efficiency, a function that depends on z_0 and z_{0s}, the smooth roughness length:

$$f_{\text{eff}} = \left[1 - \left(\frac{ln(z_0/z_{0s})}{ln(0.35[(0.1/z_{0s})^{0.8}])} \right) \right] \qquad [4.14]$$

This function makes it possible to take into account the fact that non-erodable elements (rocks) dissipate a part of wind momentum, thereby moderating saltation.

4.5.3. *Calculating emission fluxes*

Based on all the data described above, there are a large variety of schemes used to calculate the emitted fluxes. In most models, the flux will represent an average mineral species. Even if information on the minerology is available, a single calculation is carried out and the flux is then redistributed based on the percentage of each mineral. This is not an exact approach, but is much quicker in terms of computation time. Since mineral fluxes are dependent on the wind, they are very variable and must therefore be calculated in models *online* and at the highest possible time frequency. We describe below three different approaches for the calculation of these fluxes.

The scheme from **Marticorena and Bergametti (1995)** uses the equation from **White (1986)**. The single emission is distributed in a fixed manner among three aerosol modes. The basic equation used for the calculation of the saltation flux is:

$$F_h(D_p) = \frac{K \rho_{air} u_*^3}{g} \left(1 - \frac{u_*^T}{u_*} \right) \left(1 + \frac{u_*^T}{u_*} \right)^2 \qquad [4.15]$$

with u_* being the saltation shear velocity and u_*^T being the threshold shear velocity that is dependent on the diameter of the particles D_p and on z_0. We therefore have a "horizontal" flux, which will be multiplied by a coefficient to obtain the "vertical flux" (F_v), that is, the quantity that we seek to establish through the model. The coefficient α such that $F_v = \alpha F_h$ is constant and depends only on the soil type. This parameterization is therefore simple and not costly in terms of computation time.

The scheme from **Alfaro and Gomes (2001)** also uses the saltation flux from **White (1986)**, but the vertical flux is calculated from the turbulence and the size distribution depends on this. This makes the scheme more realistic but also involves greater computation time. The scheme considers the flux to have three modes of aerosol size. According to **Alfaro et al. (1998)**, these three modes have mean mass median diameter as $d_1 = 1.5 \times 10^{-6}$m, $d_2 = 6.7 \times 10^{-6}$m and $d_3 = 14.2 \times 10^{-6}$m. The cohesive kinetic energy between the grains is calculated as the wind increases, the kinetic energy increases. The first grains emitted are the largest, with the lowest cohesive energy. These are followed by finer grains. This kinetic energy is calculated as shown below:

$$e_c = \rho_p \frac{100\pi}{3} D_p^3 \left(u_*\right)^2 \qquad [4.16]$$

The value is compared to the cohesive energy for each of the three modes and the vertical sandblasting flux then becomes:

$$F_{v,m,i}(D_p) = \int_{D_p^{min}}^{D_p^{max}} \frac{\pi}{6} \rho_p \, \beta \, \frac{p_i(D_{p,i}) d_{m,i}^3}{e_i} dF_h(D_{p,i}) \qquad [4.17]$$

with i=1,N_{class} being the number of discrete intervals for the size of soil grains.

The scheme from **Kok et al. (2014)** directly calculates the vertical flux and the emission distribution only has a single mode.

$$F_d = C_d \times f_{bare} \times f_{clay} \times \frac{\rho_a \left(u_*^2 - u_{*t}^2\right)}{u_{*st}} \times \left(\frac{u_*}{u_{*t}}\right)^{C_\alpha \frac{u_{*st} - u_{*st0}}{u_{*st0}}} \qquad [4.18]$$

where f_{bare} and f_{clay} are the fractions (0:1) of barren soil and clay soil. We also have:

$$u_{*st} = u_{*t} \times \sqrt{\frac{\rho_a}{\rho_{a0}}} \qquad [4.19]$$

u_{*st0} represents u_{*st} for totally erodable soil and is chosen such that u_{*st0} = 0.16 m s^{-1}. The dimensionless coefficient C_α is such that $C_\alpha = 2.7$. The coefficient C_d represents soil erodibility such that:

$$C_d = C_{d0} \times exp\left(-C_e \frac{u_{*st} - u_{*st0}}{u_{*st0}}\right) \quad [4.20]$$

with the dimensionless constants $C_e = 2.0$ and $C_{d0} = 4.4 \times 10^{-5}$. The vertical flux is integrated over the entire size distribution and redistributed into a fixed distribution with a single mode.

$$\frac{dV_d}{dlnD_d} = \frac{D_d}{c_v}\left[1 + erf\left(\frac{ln(D_d/D_s)}{\sqrt{2}\,ln\sigma_s}\right)\right] exp\left[-\left(\frac{D_d}{\lambda}\right)^3\right] \quad [4.21]$$

with V_d being the normalized volume of mineral aerosols for each median diameter D_p, $C_v = 12.62$ μm, $\sigma_s = 3.0$, $D_s = 3.4$ μm and $\lambda = 12.0$ μm.

4.5.4. *Examples of emission fluxes*

Various examples illustrate the large variability in the emission fluxes calculated, depending on different forcing parameters or other parameterizations. The first example (Figure 4.13, left) depicts the flux as a function of the wind speed. It is calculated for the three schemes presented earlier. The differences are large and show the large variability in results for identical meteorological and surface conditions.

The second example (Figure 4.13, right) depicts a flux calculated for a wind speed of 12 m.s^{-1} and with schemes from **Alfaro and Gomes (2001)** and **Kok et al. (2014)**. The size distribution is chosen to be quite fine and this makes it possible to see the differences between the emitted fluxes. We also note a large difference between the results, depending on the parameterizations, especially in the size distribution. This will have several impacts: (i) the optical properties of the aerosols will be different, especially the calculations of optical depth; (ii) the transport over long distances will be different, especially with sedimentation (which depends nonlinearly on the aerosol size).

Mineral dust emissions are a problem whose threshold level and surface data accuracy and physics are all on a micro-scale. The entire difficulty of modeling them resides in being able to transpose turbulence calculations on centimetric scale onto meshes whose size is several kilometers. To overcome this difference in spatial scale, we must develop parameterizations while continuing to combine deterministic calculations and subgrid variability, while also taking the threshold of the emissions into account.

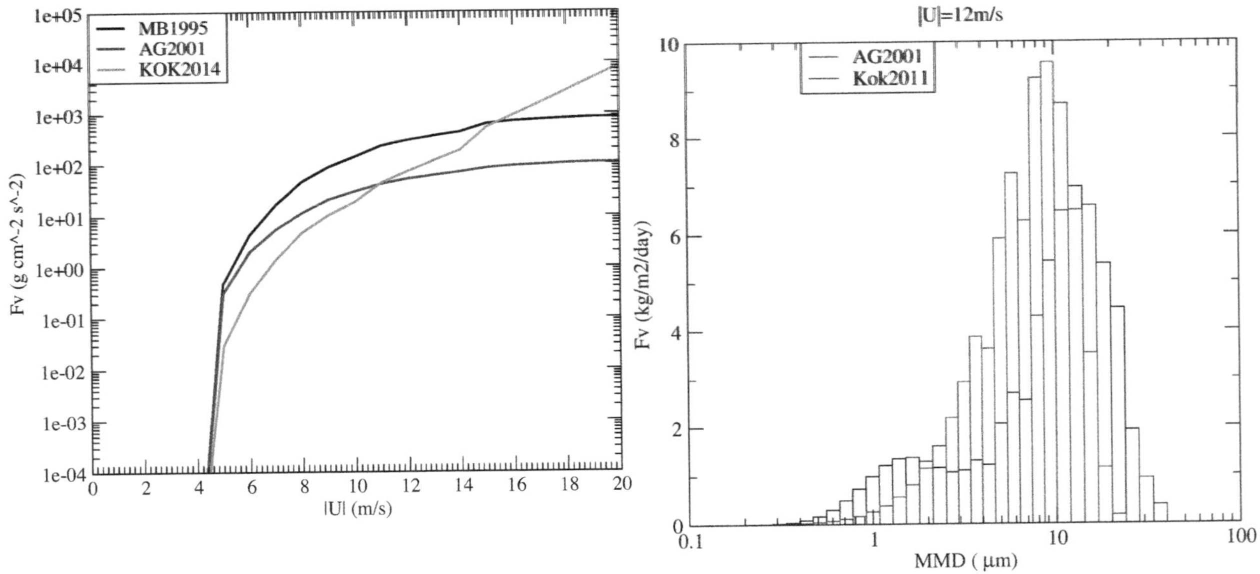

Figure 4.13. *Emission flux of mineral emissions calculated with three parameterizations (left). Comparisons of the size distributions of a flux emitted for a wind speed of 12 m.s^{-1} (right). For a color version of this figure, see: http://www.iste.co.uk/menut/modeling.zip*

4.6. Emissions from volcanoes

4.6.1. *General information*

Emissions from volcanoes are not calculated systematically for simulating atmospheric pollution, as with anthropogenic and biogenic emissions. The "exceptional" nature of these emissions leads to the fact that the studies devoted to them are generally (i) climate studies and (ii) impact studies for a very specific case. The unique feature of these emissions is that the injection of gas and particles may occur at a very high altitude, both in the troposphere as well as in the stratosphere. These emissions are mainly composed of SO_2, which may be converted into H_2SO_4 particles (Figure 4.14). When this happens, there may be a significant impact on the radiative balance, which happened during the 1991 Pinatubo eruption in the Philippines. It was estimated that there was a $-0.6°C$ drop in temperature in this latitude for three years following the eruption.

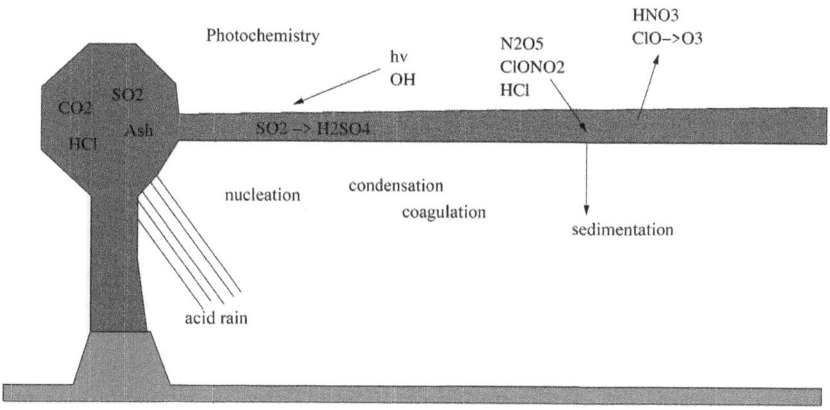

Figure 4.14. *Volcanic emissions and the associated chemical reactions in the volcanic plume. For a color version of this figure, see: http://www.iste.co.uk/menut/modeling.zip*

The orders of magnitude of these volcanic emissions and their impact on the radiative balance are summarized in Table 4.2.

Altitude	Emitted fluxes of SO_2	Radiative flux
Boundary layer	$\approx$70 Tg/year	-0.2 W/m^2
Troposphere	$\approx$10 Tg/year	-0.3 W/m^2
Tropopause	$\approx$1 Tg/year	-0.5 W/m^2

Table 4.2. *Order of magnitude of emissions and their radiative impact*

With emissions that are 70 times lower than emissions from human sources, a volcano may have twice as large an impact on the radiative balance: this is due to the altitude at which the plume stays, and also its residence time in the atmosphere. The modeling of volcanic emissions has several points in common with the modeling of emissions from forest fires: gases and particles are rapidly injected into an atmospheric column that could reach into the stratosphere, and then be transported over long distances and for long periods.

There is less uncertainty related to emissions from volcanoes than for other processes that are modeled: the source does not move, the injection height can be easily observed and the composition of a volcanic eruption is relatively well known. The only big uncertainty that remains is the emitted mass and the shape of the vertical injection profile. Recent studies have also shown that the chemical reactions in the plume are still not well understood.

4.6.2. *A recent eruption: the Icelandic volcano in 2010*

The Icelandic volcano Eyjafjallajökull erupted in April 2010 and drew a great deal of attention in Europe as it interrupted airline traffic. There were also questions about the possible fallout of the gas and particles and, therefore, the impact on health across Europe. This eruption was the subject of many studies, both through measurement and modeling.

Measurements were carried out using various methods. For example, the study carried out by *Sicard et al. (2012)* observed the plume over the Iberian peninsula using LIDAR and sun-photometer measurements from EARLINET stations. The study showed that the plume was quite fine (with a thickness of less than a kilometer) and mainly concentrated at an altitude of 11 km. It had a relatively low optical density (<0.1). The composition of the plume seemed to be mixed (ash and non-ash substances) and its highest concentration was 200 $\mu g.m^{-3}$. *Chazette et al. (2012)* also used LIDAR measurements, but this was airborne LIDAR data. This study also identified the main optical characteristics of the plume and showed that the concentration was between 140 and 250 $\mu g.m^{-3}$ at a minimum, with maxima of 1,500 and 2,300 $\mu g.m^{-3}$ depending on the location of the plume. As seen in Figure 4.15, the LIDAR data made it possible to position the volcanic plume and differentiate it from water-bearing clouds and ice. *Pappalardo et al. (2013)* complemented the LIDAR observations with an analysis of all the data from the Earlinet network above Europe for the period of the event (15 April–26 May 2010). *Waquet et al. (2014)* analyzed POLDER/PARASOL satellite data and documented the optical properties of the aerosol. They showed that close to the volcano the aerosol had a radius of 1.75 μm and an Angstrom exponent of 0. However, further away, in a plume that had already been transported, the aerosol consisted of a mixture of sulfates and volcanic dust. The Angstrom exponent here is 0.8. They noted that the particles were non-spherical.

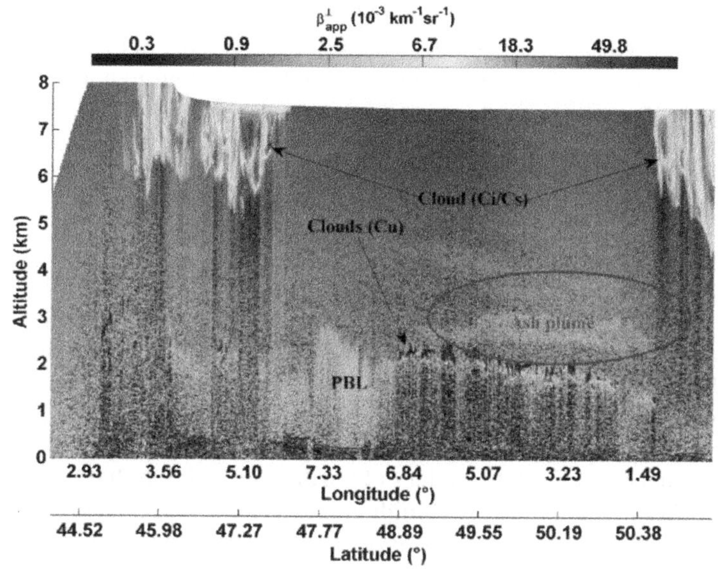

Figure 4.15. *Tracking the volcanic plume above France using airborne LIDAR data. The figure presents the backscattered signal β. Adopted from* Chazette et al. (2012). *For a color version of this figure, see: http://www.iste.co.uk/menut/modeling.zip*

To obtain more information on this plume, simulations of chemistry-transport models were performed. *Emeis et al. (2011)* carried out a simulation a few days after the eruption at the level of Germany and arrived at the observations made for the plume using the MCCM model. They showed that the aerosol layer was relatively fine and sedimented quickly. By comparing these data with surface measurements, they showed that the model was also able to reproduce the sites and the moment where the plume was integrated into the boundary layer and therefore when the particles were close to the surface. The same kind of study was carried out in France in **Colette et al. (2011)**. *Vogel et al. (2014)* carried out simulations over Germany using sets to reduce the uncertainty related to the transport of the plume. They used the following formulation for the source term:

$$Q = 100 \times \frac{h}{8500}^{1/0.241} \qquad [4.22]$$

with h being the peak of the plume at emission and 100 as the prescribed constant. This particularly simplistic formulation demonstrated how little was known regarding the emissions from volcanoes in terms of emitted mass, since this cannot be measured. Finally, the modeling made it possible to improve our knowledge of emissions. Using many scenarios and comparisons with IASI satellite measurements, *Boichu et al. (2013)* have shown that it was possible to refine the temporality of the emission period, as in reality the emitted fluxes were highly variable from hour to hour, for the entire period.

NOTE.– The articles cited on this eruption can be found in a special issue of the ACP journal: https://acp.copernicus.org/articles/special_issue212.html.

4.6.3. *An older eruption: Laki in 1783*

An example of a study on a major volcanic eruption is the study of the 1783 Laki eruption. There are, of course, very few measurements, but the eruption had a major climatic impact whose long-term consequences are still being studied. The eruption is suspected to have disrupted climatic conditions for many years, notably causing very harsh winters, followed by famines **(Thordarson and Self 2003)**, and therefore leading to higher mortality rates **(Schmidt et al. 2011)**. To understand how important this eruption may have been, studies have sought to reproduce the event. The base study is the work of **Kington (1988)**: using pressure and temperature measurements carried out on the surface in Europe, he reconstructed pressure maps. Using these pressure maps, studies like **D'Arrigo et al. (2011)** and **Yiou et al. (2014)** sought to reproduce the meteorology of this era by finding analogs. This made it possible to model the event using global meteorological ranges (since it has been shown that the year 2010 was a good analog of 1783). Based on these meteorological simulations, it is then possible to better characterize the chemistry in the plume *(Stevenson et al. 2003)* and its radiative impact *(Highwood and Stevenson 2003)*.

4.7. Emissions from lightning

Open-access publications: *Schumann and Huntrieser (2007).*

Lightning is an important source of NO_x in the troposphere. This process has been studied for many years, mainly on a global scale, and the physical and chemical fundamentals of this process have been explained in synthetic articles like *Schumann and Huntrieser (2007)*, **Price (2013)** and **Murray (2016)**. It is estimated that this source of NO_x represents 10% of all the NO_x emitted in the world, with a maximum at the equatorial latitudes *(Albrecht et al. 2016)*. The relative part of NO in the NO_x emissions is around 75–95% **(Price et al. 1997)**. The principle underlying NO_x emissions is explained in detail in **DeCaria et al. (2000)**. NO is produced when heat from the flash induces a thermal dissociation of O_2 and then a recombination with dinitrogen N_2 in a way that is similar to what happens during combustion. NO is quickly oxidized to form NO_2 and a balance is reached between NO and NO_2. The schemes in the models consider an instantaneous equilibrium between two chemical species. The impact of this production on atmospheric chemistry is given in detail in **Murray (2016)**.

Many processes must be understood and parameterized in order to quantify these emissions. (i) First of all, we must estimate the number of lightning flashes. To do this, we must establish a relation between the meteorology, the zone being studied and the flashes. (ii) It is also important to recognize the type of lightning: the emissions depend on the nature of the lightning, cloud-ground (CG) or intra-cloud (IC). (iii) We must then estimate the quantity of NO_x emitted by a flash. (iv) Finally, we must also estimate the vertical distribution of emissions along the flash (on the surface of the cloud-tops). All these points are very uncertain and there are very few observations available to constrain the calculations *(Kang et al. 2019b)*.

There are many types of parameterizations used to calculate the number of lightning flashes. These are mainly based on three approaches: (i) the "cloud top height" (CTH) scheme from **Price and Rind (1992)** is based on a linear relationship between cloud height and the frequency of lightning over the continent; (ii) the "convective precipitation" schemes from **Meijer et al. (2001)**, *Allen et al. (2012)*, *Kang et al. (2019a)*, **Kang et al. (2020)**, for example, with community multi-scale air-quality model (CMAQ), and (iii) the "ice flux" scheme by *Finney et al. (2014)*. These schemes are used in many models with some adaptations and scale factors, depending on the resolution of the models, the regions being studied and the period in the year. The uncertainty is so great that many studies have been devoted to the implementation of various schemes and the comparisons between them, as in **Clark et al. (2017)** and **Gordillo-Vázquez et al. (2019)**. These comparisons rarely conclude with a better scheme, as the number of lightning flashes is always prescribed after comparisons with observations such as satellite observations using TRMM-LIS and OTD, presented in *Beirle et al. (2010)* and *Cecil et al. (2014)*. It can be seen that these observations themselves remain very uncertain **(Marchand et al. 2019)**.

The abundance of NO_x modifies the O_3 and HO_x concentrations nonlinearly, depending on the chemical regimes **(Schumann and Huntrieser 2007)**: close to the emissions sources (like urban centers), the freshly emitted NO_x emissions (principally from NO) can reduce O_3 through titration, while for leeward emission sources, O_3 may be produced when NO_x and VOCs are present along with solar irradiation. Reactions with OH induce an increase in NO_x and cause a positive radiative forcing leading to heating via O_3. NO_x concentrations are also highly dependent on their conversion into nitric acid (HNO_3) and particulate nitrate (NO_3^-): this corresponds to the principal NO_x sinks via their dry and wet deposits. The impact of NO_x on ozone will also depend on the altitude of emissions via chemical regimes. As highlighted by **Murray (2016)**, the production of NO_x in the free troposphere will increase the concentrations of OH, NO_3^-, H_2O_2 and ozone. A greater quantity of atmospheric NO_3^- also increases its deposition. The production of lightning also has an impact on aerosols, with NO_x being oxidized with OH to nitric acid (HNO_3), and then condensed to form ammonium nitrate (NH_4NO_3) in the presence of excess ammoniac (NH_3), after neutralizing the sulfate aerosol. The net impact is a leeward increase in the aerosol due to more rapid rates of oxidation of SO_2 to sulfate and of NO_x to nitrate. It also increases the conversion of DMS to SO_2, and then to sulfate aerosol in marine environments. Regarding nitrate, complex interactions between chemistry, climate and lightning are described in *Tost (2017)*. First of all, they show that the NO_x from lightning is an important source of nitrate in the upper troposphere. It also has an impact on the formation of sulfate in the gaseous phase, then on the size distribution of the aerosols, then on the extinguishing properties of the aerosols and clouds, then on a decrease in short-wave radiation. On the other hand, certain studies describe the impact aerosols have on the production of NO_x by lightning through the direct and indirect effects of aerosols *(Wang et al. 2018)*.

4.7.1. *Frequency of lightning flashes*

4.7.1.1. *From cloud tops*

One of the most popular schemes used to calculate emissions in an atmospheric column is the scheme from **Price and Rind (1993)**. It has been implemented in various models, for example, by *Wong et al. (2013)*, *Finney et al. (2014)* and *Menut et al. (2020a)*. The prerequisite is using a deep convection diagnostic scheme to evaluate the occurrence of convective clouds. If the convective clouds are diagnosed, their upper altitude is evaluated and directly used in the parameterization to estimate the location and frequency of lightning flashes. This involves using the CTH approach:

$$\begin{cases} F_c = 3.44 \times 10^{-5} \times H^{4.9} \\ F_m = 6.40 \times 10^{-4} \times H^{1.73} \end{cases} \quad [4.23]$$

with F_c and F_m being the frequencies of lightning (# mn^{-1} 25 km^{-2}) *(Banerjee et al. 2014)*, with c used for continental and m for maritime. H is the top of the convective cloud (km). A scale factor **(Price and Rind 1994)** designed to take into account the resolution of the model is applied:

$$c_{pr} = 0,97241 \ \exp(0.048203 \times \Delta\lambda \times \Delta\Phi) \quad [4.24]$$

where $\Delta\lambda$ and $\Delta\Phi$ are the dimensions of the cell in latitude and longitude, respectively, expressed in degrees. In practice, except for very coarse resolutions, this factor evolves around 0.97 and, in practice, has a very small impact on the results, compared to the uncertainties of all the other parameters. These uncertainties are generally compensated for by adjustment factors, to get closer to observation, as described in **Gordillo-Vázquez et al. (2019)**. This kind of CTH scheme has been extensively tested (e.g. *Wong et al. (2013)*). For instance, *Finney et al. (2014)* used an additional scale factor of $\alpha = 0.05$ to adjust the number of lightning flashes after comparison with global annual satellite data. For each cell in the model, the relative percentage over sea, x_{sea} is estimated using the land-sea mask for landuse databases. The frequency of the lightning is then estimated as follows:

$$F = c_{pr} \times (x_{sea} \times F_m + (1 - x_{sea}) \times F_c) \qquad [4.25]$$

with F being the final number of lightning flashes, divided by the area of the cell, to obtain the number per km^2, therefore in (# mn^{-1} km^2). For each lightning flash, we must estimate the relative part that is in the cloud (IC) and from the cloud to the ground (CG). This estimation is principally based on the estimation of the cloud ice depth, H_{fr}, proposed by **Price and Rind (1993)** and calculated as follows:

$$H_{fr}(km) = -6.64 \times 10^{-5}|L|^2 - 4.73 \times 10^{-3}|L| + 7.34 \qquad [4.26]$$

with L being the latitude, in degrees. Note that the adjustment has only been given for the Northern Hemisphere. For a symmetric relation between both hemispheres, the absolute value of the latitude is used. H_{fr} is used to calculate the parameter β:

$$\beta = 0.021H_{fr}^4 - 0.648H_{fr}^3 + 7.493H_{fr}^2 - 36.54H_{fr} + 63.09 \qquad [4.27]$$

p is defined as the relative part of CG in the total (IC+CG) and is defined as $p = 1/(1+\beta)$. It must be noted that β is used with limited values and must be $1 < \beta < 50$. This corresponds to 5.5 km $<$ H_{fr} $<$ 14 km, which makes it possible to avoid unrealistic values.

4.7.1.2. *From convective precipitations*

Meijer et al. (2001) propose a parameterization based on convective precipitation. This scheme is calibrated using correlations with meteorological variables calculated by the ECMWF model and compared with measurements from the EULINOX campaign in Europe during the summer of 1998. The hypothesis is used to distinguish convective clouds: all clouds whose base is lower than 700 hPa and whose thickness is smaller than 5 km are considered to be convective. The statistical analysis of these clouds finally yields the relation:

$$F_c = 14,700 \times CP + 1.7 \qquad [4.28]$$

where F_c is the number of flashes and CP is the convective precipitation (in mm or m, according to **Finney et al. (2014)**). They then considered the constant value of the energy of a flash: 10^{16} molecules NO/J. NO production takes place during the phase where the flash returns from the cloud to the soil. On average, the energy deployed is 6.7 × 10^8J. **Allen et al. (2009)** use the same principle in the CMAQ model. The production is calculated such that:

$$LNO_x = k \times LF \times E \times P \qquad [4.29]$$

with LF being the frequency of the flashes, E being the energy of a flash, P being the production of NO per unit energy and k being a conversion factor such that $k = M_w(NO)/Avog$ (M_w is the molecular mass of nitrogen, N). Unlike other schemes, this study considers that $E_{IC} = E_{CG}$. They use the constant $P = 500$ moles N per flash. We must then estimate the frequency of the flashes for each grid cell in the model:

$$LF(i, j) = G \times \alpha(i, j) \times (P_{conv} - T_{prec}) \qquad [4.30]$$

with P_{conv} being the rate of convective precipitation (mm/h), G and $\alpha(i, j)$ being adjustment factors (on the global and regional scale, respectively), and T_{prec} being the threshold value beyond which there is no flash (unknown value). While the scheme in **Allen et al. (2009)** is simple in principle, there is not enough information in the article to be able to use it. Notably, they use observations from the mNLDN (monthly National Lightning Detection Network) network in the United States to calibrate the constants. This scheme is used by *Appel et al. (2011)* in CMAQ; however, the article also lacks enough detailed information to be able to use it. **Allen et al. (2012)** return to the scheme, with added details: the threshold is set to 0, the rate of convective precipitation is only considered for a pressure below 450 hPa. There is no information on the other adjustment constants, which are estimated using measurements from the United States carried out over specific periods. The results show that without "local scaling" or subgrid adjustment using radar measurements, results that are based only on the height of the convective cloud-top are quite biased. While the scheme from **Meijer et al. (2001)** is simple and usable as it is, the scheme by **Allen et al. (2009)** and **Allen et al. (2012)** is not detailed enough in the articles to be usable.

4.7.1.3. *Using ice flux*

Finney et al. (2014) proposes an approach based on the upward flux of ice. Their scheme has been calibrated using ECMWF simulations. As with the scheme from **Allen and Pickering (2002)**, they considered, as in the article by **Allen and Pickering (2002)**, that the pressure level 440 hPa is representative of vertical movements throughout the cloud. In their scheme, the flux of the mass of upward ice ϕ_{ice} ($kg_{ice}m_{cloud}^{-2}s^{-1}$) is estimated at this altitude using the relation:

$$\phi_{ice} = \frac{q \times \Phi_{mass}}{c} \qquad [4.31]$$

with the variables at 440 hPa, q is the specific cloud ice water content $(kg_{ice}kg_{air}^{-1})$; Φ_{mass} is the updraught mass flux $(kg_{air}m_{cell}^{-2}s^{-1})$; c is the fractional cloud cover $(m_{cloud}^2 m_{cell}^{-2})$. The flux ϕ_{ice} takes the value zero when $c < 0.01 m_{cloud}^2 m_{cell}^{-2}$. The relationship between the ice flux and the flashes is established based on LRMTS data (LIS-OTD low-resolution monthly time series). This makes it possible to carry out a regression between the two and obtain:

$$\begin{cases} f_l = 6.58 \times 10^{-7}\ \phi_{ice} \\ f_o = 9.08 \times 10^{-8}\ \phi_{ice} \end{cases} \qquad [4.32]$$

with f_l and f_o being the density of the flashes $(fl.m_{cell}^{-2}s^{-1})$ above the land and the ocean.

4.7.1.4. *Cloud resolving models*

Barthe and Barth (2008) proposed a scheme dedicated to resolving clouds (*cloud resolving models* (CRM)) and implemented this in the WRF. This scheme differentiates between whether or not there is icy precipitation when calculating the flash rate:

$$F_{MF} = 1.13 \times 10^{-15} \times f_{np} \times f_p \qquad [4.33]$$

with F_{MF} being the flash rate (fl.min^{-1}) calculated from the mass of ice-flux, both precipitating and non-precipitating: f_p (kg m s^{-1}) and non-precipitating f_{np} (kg s^{-1}). The constant (1.13) has been determined through adjustment with data from the STERAO campaign. *Gressent et al. (2016)* implemented a plume-in-grid, at high-resolution, within the Meso-NH to better account for the chemistry specific to a massive injection of NO_x in the upper troposphere. They show that the subgrid variability induces a variation of $\pm 20\%$ for NO_x and $\pm 5\%$ for ozone.

4.7.2. *Estimation of NO production*

4.7.2.1. *Estimation of the emitted masses*

Schumann and Huntrieser (2007) published a review and showed that the annual, global NO_x from lightning ranged from 3 to 20 Tg(N) year^{-1}. They also showed that a typical flash emits 15 (2-40) $\times$ 10^{25} molecules of NO. Finally, they showed that production from lightning represented $\approx$10% of the global production, of which $\approx$4% was produced between latitude 35°N–60 °N. **DeCaria et al. (2000)** calculated the production of NO based on the size of the lightning strike, the atmospheric pressure and various constants evaluated in laboratory experiments. **Price et al. (1997)** proposed the following to relate the frequency of flashes to an NO_x concentration:

$$\begin{cases} P(CG) = 6.7 \times 10^{26} \text{ molecules NO} \\ P(IC) \;= 6.7 \times 10^{25} \text{ molecules NO} \end{cases} \qquad [4.34]$$

According to *Mari et al. (2006)*, these values are high compared to other publications on this subject. Finally, the production term for updrafts is:

$$\frac{\partial NO}{\partial t} = c_{pr}\left(P(IC) \times \frac{\beta\, f_f}{1+\beta} + P(CG) \times \frac{f_f}{1+\beta}\right) \qquad [4.35]$$

Barthe and Barth (2008) returns to the scheme given by **Wang et al. (1998)** with:

$$n_{NO}(P) = a + b \times P \qquad [4.36]$$

where n_{NO} is the number of NO molecules produced per flash (molecules m^{-1}), P is the pressure (Pa), $a = 0.34 \times 10^{21}$ and $b = 1.3 \times 10^{21}$. *Barret et al. (2010)* inter-compared models from the region covered by the *African Monsoon Multidisciplinary Analysis* (AMMA) project (Lebel et al. 2010). Four models were involved, using different parameters. However, it can be noted that at its core each parameterization is always based on either **(Price and Rind 1992)** (cloud-top-height) or **(Meijer et al. 2001)** (convective precipitation). Ozone profiles are compared using MOZAIC data. They concluded that the models qualitatively reproduced observations, and that there were large differences seen between the models.

4.7.3. *The relative part produced by CG and IC*

NO_x production differs greatly depending on whether the lightning remains in the cloud (IC) or goes from the cloud toward the ground (CG). A polynomial relationship by **Price and Rind (1993)**, relating the ice cloud thickness (H_{fr} (km)) to the ratio between IC and CG is used in many studies, including *(Wong et al. 2013)* such that:

$$\beta(\%) = \frac{CG}{IC + CG} = 0.021H_{fr}^4 - 0.648H_{fr}^3 + 7.493H_{fr}^2 - 36.54H_{fr} + 63.09 \qquad [4.37]$$

with β being a bounded parameter such that: $1 < \beta < 50$. This leads to the thickness as it was observed: $5.5\,\text{km} < H_{fr} < 14\,\text{km}$. The thickness ($H_{fr}$) has also been determined from estimations based on measurements, so that:

$$H_{fr}(km) = -6.64 \times 10^{-5}L^2 - 4.73 \times 10^{-3}L + 7.34 \qquad [4.38]$$

with L being the latitude in degrees. On the contrary, *Finney et al. (2016)* considered that there is no difference between IC and CG. According to them, the relationship between the number of flashes and the NO_x emissions therefore means that (i) a flash produces 1.2 GJ and (ii) the production of NO is 12.6 $\times 10^{16}$ NO molecules J^{-1}.

4.7.4. *The vertical emission profile*

The vertical emission profile seems to be difficult to quantify, but the results are very sensitive to this **(DeCaria et al. 2000)** since: (i) the NO_x have longer lifetime in the upper troposphere (UT), (ii) the production efficiency of ozone is greater in the UT. **DeCaria et al. (2000)** used single-mode distributions (CG) and two-mode distributions (IC). *Schumann and Huntrieser (2007)* compared different approaches to determine the shape of the vertical profile. The most widely used shape is the "C-shaped profile" which corresponds to the profile parameterized by **Pickering et al. (1998)**: two peaks, one in the lower troposphere and one in the upper troposphere. However, **Ott et al. (2010)** believe that this shape places too many emissions close to the surface and not enough in the troposphere: they suggest a vertical profile with low values in the lower troposphere. **Mari et al. (2006)** include the calculation of the shape of the vertical profile of the NO_x in the deep convection scheme in the MEso-NH. The approach is validated at a small spatial-temporal scale by comparing calculations of airborne measurements from the TROCCINOX campaign (Bauru, tropical latitude, March 2004). In their study, the value of the NO_x products is the same as in **Price and Rind (1992)**, but with refinements from the same authors.

4.7.5. *Intercomparisons and observations*

4.7.5.1. *Intercomparisons of parameterizations*

Tost et al. (2007) implemented different schemes in the ECHAM5 model. They concluded that no scheme is better than any other when compared with observations. They showed that each scheme is made up of so many adjusted constants that in its entirety this makes all the schemes very uncertain. **Finney et al. (2014)** intercompared various known schemes with his new scheme based on ice flux. A study of the same type by **Luo et al. (2017)**, compared emission schemes as well as vertical profiles. They concluded that the most robust parameterization remains that used by **Price and Rind (1992)**, as its uses robust dynamic variables, while the others use more uncertain parameterization results (convective precipitation, updraft mass flux, microphysics diagnostics).

4.7.5.2. *Estimating NO_x production based on observations*

Two types of observations are commonly used:

– data from surface networks, such as the National Lightning Detection Network (NLDN) in the United States. Only CG flashes are detected;

– satellite data: it is not possible to distinguish between IC and CG.

Boersma et al. (2005) inverted the NO_2 profiles from the GOME satellite spectrometer to quantify NO_x production correlated with lightning. From these measurements, they confirmed the choices made by **Price and Rind (1992)**, as well as the local maximum production on convective cloud-tops. They found there was a ratio of 1.6/1 for convective cloud production (for the same value of H) between the land and the sea. They tested two different parameterizations to calculate production from lightning and compared the parameterization results with the GOME profiles. After comparisons with the observations, they proposed correction coefficients for the parameterization. However, these coefficients are highly dependent on the period and location in which the study is carried out and cannot really be generalized. *Höller et al. (2009)* used measurements from a LINET system (lightning detection network) to quantify the frequency of lightning in Brazil, Africa and Australia. *Beirle et al. (2010)* used SCIAMACHY satellite data to quantify the efficiency of NO_x production through lightning. They found that previous estimates had overstimated this with respect to their results, notably because (i) the published articles had principally studied cases of high production in order to be illustrative, and (ii) most studies were carried out in the United States and were not considered to be representative at a global level. *Peterson and Beasley (2011)* studied the potential effect of NO_x catalysis when lightning passed through ice-crystals. They believed that taking this mechanism into account would increase NO_x production, especially in IC flashes, and would thus shrink the gap between IC and CG. *Kaynak et al. (2008)* directly used these data in the US National Lightning Detection Network database and simulated this lightning with MM5 and CMAQ. This database only covers the United States. The results for ozone, over two summer months, showed very small differences with respect to simulations with and without lightning. *Miyazaki et al. (2014)* assimilated OMI, MLS and MOPITT data in a CTM to adjust the NO_x flux. This flux is estimated using the scheme by **Price and Rind (1993)**. They arrived at an annual mass that was smaller than previous estimates (6.3 TgN year^{-1}). They also showed that satellites could improve the vertical profile of the emitted NO_x and that the commonly used schemes over-estimated the altitude of the profile peak by 1 km on average.

Globally, the calculation of these emissions remains very uncertain and the use of data can only bring in fragmented information. Even if these emissions are not on the surface, they play a key role in understanding atmospheric chemistry and are among the most uncertain quantities at present. Using measurements, it is possible to approximately locate lightning flashes (but not if there is heavy cloud cover) and locate the maximum cloud altitude. On the other hand, it is not possible to measure the emitted flux or the shape of the vertical profile.

4.8. Natural marine emissions

Open-access publications: *Gantt and Meskhidze (2013), Grythe et al. (2014).*

4.8.1. *Sea-salts*

Marine emissions are largely sea-salt emissions. These are aerosols that are chemically composed of Na and Cl and which are stripped away from the ocean surface by wind friction. Emitted in large quantities (although the global mass is still quite unknown), they play an important role in the earth's radiative balance **(Lundgren et al. 2013)**. However, they are not the most harmful in terms of health and atmospheric pollution. On the contrary, in all coastal regions, they may be advected in great quantities over land and must therefore be considered when modeling pollution, as they become part of the particle mass being studied **(Im 2013)**.

The process by which sea spray aerosols (SSA) are emitted is described in the article by *Grythe et al. (2014)*. Their size may vary between a mean mass median diameter (MMD) of 10 nm and a few millimeters. Their residence time in the atmosphere can range from a few seconds to a few days. The determining factor for emission is wind close to the surface: this provokes the formation of waves and the emission of the aerosols is induced from their crest or *spume*. This is called a "direct effect". Another phenomenon is related to the bursting of bubbles, which also releases emissions, called an indirect effect. It was first shown that the fluxes were proportional to the wind speed. The earliest schemes only showed this dependency. More recently, it was shown that there was also a dependence on the sea surface temperature (SST). Finally, the most recent schemes also take into account the salinity, as this can influence the density of the crests of the wave, as well as the chemical composition of the aerosols emitted. The most widely used emission scheme is probably the scheme from **Monahan (1986)**, which only depends on the wind:

$$\frac{dF}{dr} = \frac{1.373 \times U_{10}^{3.41}}{r^3} \left(1 + 0.057 r^{1.05}\right) 10^{1.19 e^{-B^2}} \quad [4.39]$$

$$\text{with } B = \frac{0.38 - log(r)}{0.65} \quad [4.40]$$

Here, F is the emitted flux (particles m^{-2} s^{-1} μm^{-1}), r is the radius of the particle (μm) and U_{10} is the wind at 10 m (m s^{-1}). The flux is calculated with a few limitations: the size distribution is fixed and the flux is only calculated for $U_{10} < 30ms^{-1}$. As **Grini et al. (2002)** noted, this process only represents the indirect mechanism of the bubbles bursting. Originally, this scheme was only intended for aerosols with $D_p > 0.8\mu m$. **Martensson et al. (2003)** and *Grythe et al. (2014)* have also proposed other schemes to represent fine particles. These schemes can also calculate a distribution with variable size. The scheme by **Martensson et al. (2003)** is, however, limited for $D_p < 2.8\mu m$. The scheme proposed by *Grythe et al. (2014)* is:

$$\begin{aligned}\frac{dF(D_p,U_{10},T)}{dD_p} &= T_w.\left[235.U_{10}^{3.5}exp\left(-0.55\left[ln\left(\frac{D_p}{0.1}\right)\right]^2\right)\right.\\&+0.2U_{10}^{3.5}exp\left(-1.5\left[ln\left(\frac{D_p}{2}\right)\right]^2\right)\\&\left.+6.8U_{10}^{3.0}exp\left(-1.0\left[ln\left(\frac{D_p}{30}\right)\right]^2\right)\right]\end{aligned} \quad [4.41]$$

with T_w (K) being the temperature dependence on the aerosol production, for a given temperature T, such that:

$$T_w(T) = 0.3 + 0.1\,T - 0.0076\,T^2 + 0.00021\,T^3 \quad [4.42]$$

The hypothesis used for the calculation is for a relative humidity of 80%, signifying that the damp radius is twice as large as the dry radius of the aerosol. For these emission schemes, the process is largely dynamic and very little input data are required. The information we require is the land-sea mask, which is usually already known in the simulation domain. This information is still required in the form of a map or coupling with an ocean model, even in the most recent schemes, which include the influence of salinity. From a calculate flux value, we then need to divide the number of aerosols into chemical species **(van den Berg et al. 2000)**. According to **Seinfeld and Pandis (2006)**, the chemical speciation is of the order: 30.6% sodium (Na^+), 55.04% chlorine (Cl^-) and 14.4% other, inorganic compounds. *Grythe et al. (2014)* give a detailed information on this speciation, with 55% Cl^-, 31% Na^+, 8% SO_4^{2-}, 4% Mg^{2+}, 1% Ca^{2+} and 1% K^+.

4.8.2. *Dimethylsulfide*

Dimethylsulfide (DMS) is an oceanic source of atmospheric SO_2. It has an impact on cloud formation and can play the role of a condensation kernel. The emitted flux is parameterized by considering that it corresponds to a fraction of the aqueous DMS present at the surface of oceans. Thus, to calculate this flux we must have a map of surface concentrations, like that created by **Lana et al. (2011)**. This global map has a monthly average and a 1 degree × 1 degree resolution. The temporal and latitude dependency is related to aqueous DMS production through plankton activity, which itself depends on temperature and sunlight. According to **Zavarsky et al. (2018)**, the flux calculation can be written as:

$$F = -k.\left(\frac{C_{air}}{H} - C_{water}\right) \quad [4.43]$$

with k (m s^{-1}) being a transfer velocity, and c_{air} and c_{water} being the DMS concentrations in air (gaseous phase) and the ocean (liquid phase). H is Henry's constant, representing the solubility of the gas, and depends on the SST and salinity. A simplifying hypothesis that is often used is to consider that the concentration in water is much higher than the concentration in air, leading to:

$$F = k.c_{water} \qquad [4.44]$$

Different schemes have been developed to calculate the transfer velocity k, such as those by **Liss and Merlivat (1986)**, **Wanninkhof (1992)** and **Nightingale et al. (2000)**. More recently, **Elliott (2009)** created a composite of these schemes using relationships from **Kettle and Andreae (2000)**, which associated wind and the SST through the calculation of the Schmidt number for DMS. Once the DMS is emitted into the atmosphere, it will react chemically by being oxidized in the day by the OH radical and at night by the NO_3 radical to produce SO_2 and then SO_4^{2-}.

In the context of studying regional pollution, natural marine emissions are the second-order process. The concentrations are relatively large near the coasts, but they are not among the most intense emissions. Further, since their composition does not pose any significant health risk, they are not a priority in studies for forecasting or analyzing regional atmospheric pollution.

4.9. Pollen

Open-access publication: *Sofiev et al. (2013a).*

Pollen is a good example of an aerosol-type pollutant, which is not very well researched, although its concentration has increased greatly in recent years. From a health point of view, this has the unique feature of being easily identifiable as it causes recognizable allergies early into its emission and transport. The modeling of pollen makes use of the broad principles discussed for chemistry-transport, but with more processes to be considered and greater uncertainty for each of these processes. This is currently another domain where very little is known and is a challenge for chemistry-transport modeling. "Pollens" are a family of grains with very different natures and taxons. At present, there are four main taxons that are modeled as being highly allergenic: birch, olive, ragweed and grass.

Figure 4.16 illustrates the various processes in the lifecycle of pollen. The first challenge is related to **emissions**: depending on the type of pollen, the emissions will be from plants close to the ground or from trees. Thus, the model can have an emission injection height of between 0 and 15 m. Each pollen species is emitted by a plant with its own lifecycle. The difficulty lies in detecting the onset of pollination and its duration. The start of the period is often calculated based on the cumulative temperature of the preceding months. Unlike pollution, in order to obtain an accurate simulation of a pollen season, the *spin-up* time is initiated well in advance. Typically, for a season beginning in March, the simulation must begin in the preceding October. We must also estimate the quantity of grains that will be emitted. For certain types of pollen, like Ambrosia, the emitting plant is a perennial plant, that is, it blooms in all seasons. It is difficult to establish the emission cadastres: there are no methods for remote detection to inventory the plants. Further, as perennials spread and grow, each cadastre becomes obsolete the following year. The second challenge is to **model transport**: unlike polluting aerosols, pollen grains are larger but less dense, therefore lighter. However, the chemistry-transport hypotheses related to advection remain valid a priori (**Sofiev et al. 2006**). When the aerosols are deposited on the ground, the percentage of their possible resuspension is low. This is not the case for pollen: being dry and light, they can easily return to the atmosphere. However, this resuspension also depends on their aging: older grains get desiccated, becoming smaller and denser and thus getting resuspended less easily. The third challenge is **pollen's allergenic property**. Depending on the type of plant, the pollen may be more or less allergenic. This allergenic property will also decrease as the grain ages. Finally, it seems the same plants emit allergy-inducing grains in some years, but not others.

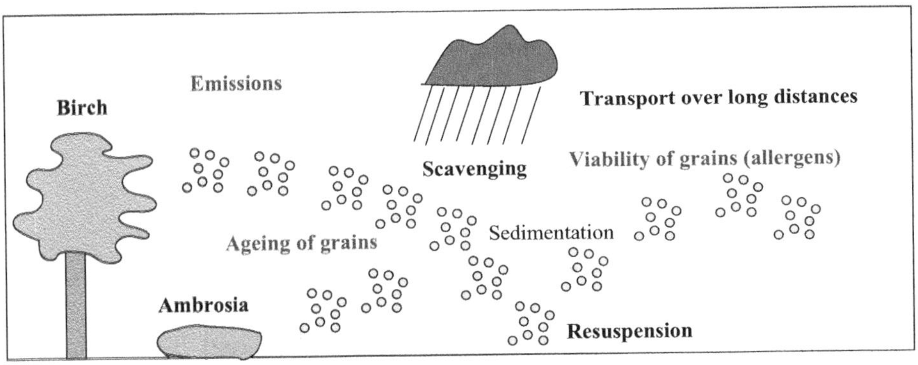

Figure 4.16. *General principle underlying the modeling of pollen. For a color version of this figure, see: http://www.iste.co.uk/menut/modeling.zip*

The emission of pollen grains is the most difficult step in this modeling: these emissions depend on the meteorology and also on the plant phenology. So in addition to the atmosphere, we must add a vegetation dimension to the model. Studies have shown that CO_2 and O_3 concentrations may have an impact on the emissions (**Rogers et al. 2006; Darbah et al. 2008**), but this has not yet been accounted for in the models. To describe the functions used in most models, we can take the example of the parameterization used by **Sofiev et al. (2013a)**, implemented in the SILAM model for birch and olive. To calculate the emissions, we must first estimate the start and end of the pollination season for the taxon being studied. The estimation of the pollination start date is based on the concept of calculating the cumulative temperature until it reaches a threshold – this is the *heat-sum concept*. This sum (H) is calculated from the 60th day of the year onwards. For each taxon and each grid cell in the model, we define a map of the threshold value (H_{start}), which will define the start of the season. This value must be spatialized as it seems to vary with latitude. The pollen emission rate, $R(t)$, increases with the sum (H) and with time (t). The end of the emission season is deduced either by calculating an average duration for the period or by defining a quantity to be emitted (N_{tot}), representing the *catkins* content, and letting the emission continue until this is completely released. Finally, this mass must be redistributed on an hourly basis in order to be able to estimate the surface flux in (grains.m^{-2}.h^{-1}). It will also be noted that the emission begins not only if $H > H_{start}$, but also if the temperature exceeds a pre-established critical value $T > T_{co}$ with dry meteorological conditions, that is, without precipitation and with low humidity. The temperature T_{co} is defined as the cutoff temperature, between 0 and 3°C, depending on the published studies. For certain models, we also take into account factors that may "accelerate" the start date. These include the wind speed and turbulence, which mechanically strip away the pollen. Finally, the emission flux calculation is expressed by:

$$E(t,i,j) = N_{tot}.\phi(i,j).\frac{T - T_{co}}{\Delta H}.f_{wind}(U,w^*).F_{in}\left(\frac{H}{H_{start}}\right)$$

$$. \ F_{out}\left(\frac{R(t)}{N_{tot}}\right).f_{th}(q).f_{th}(P) \qquad [4.45]$$

with $E(t,i,j)$ being the emitted flux in (grains m^{-2} h^{-1}) for a given time and a model grid (i,j), N_{tot} being the total number of grains emitted over 1 m^2 during the whole season, ϕ(i,j) being the fraction of area of emitting plants contained in one grid cell in the model (i,j), $(T - T_{co})/\Delta H$ being the intensity of the emitted flux, which depends on the temperature; F_{in} being the fraction of the flux emitted, which depends on the progress of the season; F_{out} being the fraction of the emitted flux, which depends on what has already been emitted with respect to the total available; f_{wind} being a weighting function related to the wind; and f_{th} being a weighting function related to humidity and precipitation. The functions f_{wind} and f_{th} take values between 0 and 1. The function f_{wind} describes an increase in emissions related to a strong wind, with:

$$f_{wind} = f_{windmax} - exp\frac{-(U + w_*)}{U_s} \qquad [4.46]$$

where U is the wind speed at the altitude of the emission, w_* is the convective vertical velocity, $f_{windmax}$ represents the wind impact intensity (for birch, $f_{windmax}$ = 1.5 m s^{-1}), and U_s represents the maximum wind value that can have an impact on emission (for birch, $U_s = 5$ m s^{-1}). The function f_{th} is a limiting fraction of emission, with:

$$f_{th} = \begin{cases} 1 & \text{if } x \leqslant x_{min} \\ \dfrac{x_{max} - x}{x_{max} - x_{min}} & \text{if } x_{min} < x << x_{max} \\ 0 & \text{if } x \geqslant x_{max} \end{cases} \quad [4.47]$$

where x represents the relative humidity q (%) or the precipitation P (mm h^{-1}). The values used to define these linear functions are $q_{min} = 50\%$, $q_{max} = 80\%$, $P_{min} = 0$ and $P_{max} = 0.5$ mm h^{-1}. There are other formulations that are used to calculate ambrosia emissions (e.g. **Makra et al. (2004)**, **Helbig et al. (2004)**, **Efstathiou et al. (2011)**, **Prank et al. (2013)** and **Zink et al. (2013)**). Taking the example of the parameterization given by **Efstathiou et al. (2011)**, we have:

$$F = a.K_e \frac{P}{LAI.H} u_* \quad [4.48]$$

where $a = 5 \times 10^4$ is an adjustment constant, K_e is a factor that varies over time and represents a limiting factor due to meteorology, P is the daily production of grains per m^{-2}, LAI is the leaf area index (chosen here to be constant, with LAI = 3 m^2 m^{-2}) and H is the height of the plant (also chosen to be constant, with $H = 1$ m).

These emission fluxes remain very uncertain. Most current models adjust the calculated flux based on observations made in previous years and by calibrating the flux using measurements. However, this research area is currently evolving significantly, as the health impacts of these emissions are quite identifiable, rapid and aggressive (allergies and asthma). Certain physical processes still remain unknown for the most part, namely, the resuspension of pollen grains, the aging and desiccation of the grain, the allergenic capacity of the grain depending on time and transport. Finally, the transport model itself may be brought under scrutiny: the hypothesis used for polluting gases and aerosols may perhaps not extend to this case. Can it be considered that the particle is still small enough to apply the hypothesis that it is displaced strictly at flow velocity?

4.10. Summarizing information on emissions

The uncertainties in estimating emissions differ over time, space and depend on the emitted pollutant. To summarize the various uncertainties discussed in this chapter:

– There is no emission that is more uncertain than others. Emissions are so variable that the uncertainty associated with a flux will depend not only the method used for estimation, but also on the spatial and temporal resolution of the database that is sought to be established.

– The nature of the emissions result in different estimation methodologies: anthropogenic emissions will primarily be diagnosed based on databases that record sources one by one, while natural emissions are estimated based on parameterizations, which are themselves based on databases representing the ground and surface.

– Thus, in order to calculate emissions we must already have a model, even before using a CTM to calculate concentration. This then multiplies the uncertainties. Knowing that we can ultimately only compare atmospheric concentrations, we must remember that the emission cadastres that we use will depend not only on the emission model, but also on the CTM (and its resolution) used to qualify the pertinence of the emitted fluxes.

5

Deposition

The chemical species being modeled may be removed from the atmospheric budget by dry or wet deposition. It is very important to represent both these processes, but modeling studies have not focused on them as much as on emissions or transport, for example.

5.1. Dry deposition

Open-access publication: *Pleim and Ran (2011).*

The parameterization of dry deposition illustrates the need to have dynamic turbulent parameters available. The principle for dry deposition is to calculate the sedimentation rate for gases or particles. In order to quantify this deposition, the most conventional parameterization **(Wesely 1989)** uses the idea of electrical resistances arranged in series. As the fall velocity acts from top to bottom, only the vertical flux components come into play in the parameterization. The lower atmosphere can be depicted on a vertical axis, divided into overlapping layers (see Figure 5.1).

For each vertical level, we can create a resistance that depends on parameters associated with the state of the atmosphere (stability or instability), the roughness of the surface and the type of vegetation. If all the resistances that a particle successively encounters are low, the particle falls and is deposited. If one of the resistances is high, then the particle's descent stops and it is not deposited. The contribution of this effect to the concentration (c) of a particle will thus act only along the vertical component, such that:

$$\Delta c = \frac{\partial}{\partial z}(v_d.c) \qquad [5.1]$$

where v_d is the deposition velocity. The deposition of a species (gas or aerosol) occurs due to three different phenomena:

Modeling of Regional Atmospheric Pollution,
Laurent MENUT. © ISTE Ltd 2024.

– Turbulent diffusion in the surface layer may lead this particle toward the ground. We must therefore know the aerodynamic resistance r_a.

– A fine, diffusing surface layer, called the laminar layer, is very thin (of the order of 0.1 cm), but plays a substantial role here. We must therefore calculate the surface resistance r_b. In the case of aerosols, we must also determine the rate of sedimentation (v_s).

– For gases, the solubility of the chemical species in this laminar layer lead to the determination of the in-canopy transfer resistance (r_c).

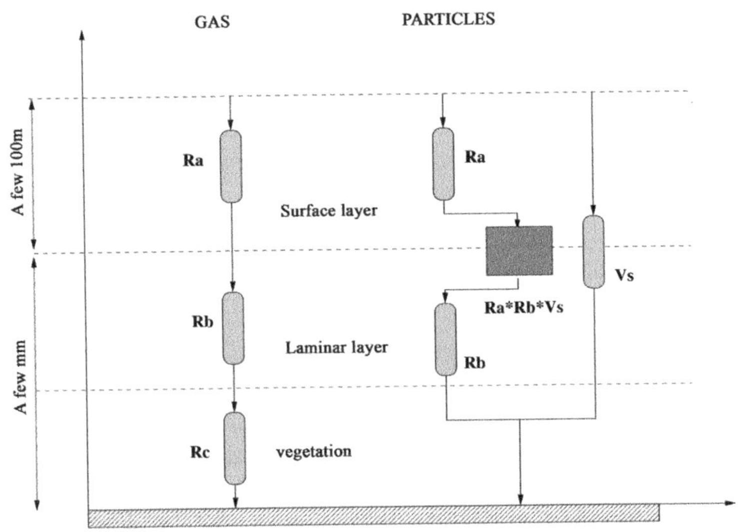

Figure 5.1. *General principle underlying the scheme used to calculate the dry deposition of gases and particles*

For gaseous species, the deposition velocity (m.s^{-1}) can be expressed as:

$$v_d = \frac{1}{r_a + r_b + r_c} \qquad [5.2]$$

For aerosols, according to **Wesely (1989)**, v_d will be:

$$v_d = v_s + \frac{1}{r_a + r_b + r_a r_b v_s} \qquad [5.3]$$

and according to an alternative approach proposed by **Zhang et al. (2001)**:

$$v_d = v_s + \frac{1}{r_a + r_s} \qquad [5.4]$$

with r_s being the surface resistance, which depends on vegetation. Figure 5.2 depicts an example of the calculation of the deposition velocity for aerosols. It can be seen that particles with a diameter of 0.2 μm have the lowest velocity.

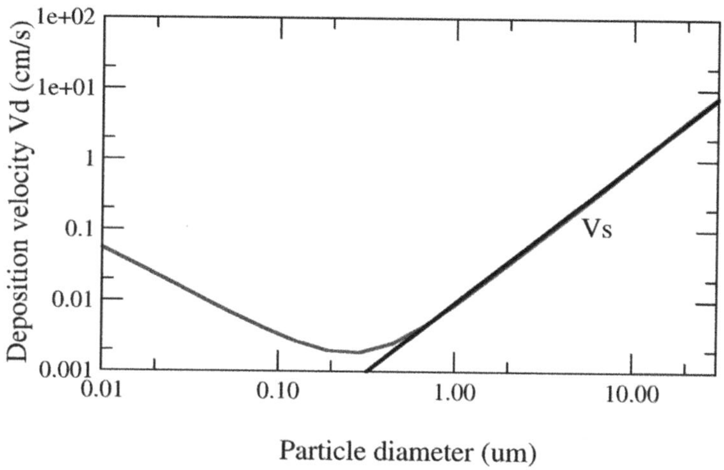

Figure 5.2. *Velocity of dry deposition (in red) calculated for particles of different sizes used in the model. The sedimentation rate is the part of the curve that is black. For a color version of this figure, see: http://www.iste.co.uk/menut/modeling.zip*

5.1.1. *Aerodynamic resistance* r_a

Aerodynamic resistance (r_a), which is common to gases and particles, reflects the turbulence close to the surface. The greater the turbulence, the less likely it is that the particle will sediment. This resistance depends on meteorological parameters such as atmospheric stability, via the Monin-Obukhov length L, the shear velocity u_*, the dynamic roughness length z_{0m} and the wind speed $|U|$:

$$r_a = \frac{1}{C_D |U|} \qquad [5.5]$$

with C_D being the coefficient of aerodynamic resistance to the transfer of momentum, such that:

$$C_D = \frac{k^2}{\left[ln\left(\frac{z}{z_{0m}}\right) - \Psi_M\left(\frac{z}{L}\right)\right]^2} \qquad [5.6]$$

and where Ψ_M is the stability function, depending on atmospheric stability in the surface layer.

5.1.2. Surface resistance r_b

Surface resistance is expressed differently for gases and aerosols. For gaseous species, r_b depends on the shear velocity u_*, the roughness length of the pollutant being studied (z_c) and the momentum roughness length (z_{0m}). This resistance will be a mixture of the impact of turbulence and the characteristic of the gas or particle being considered.

$$r_b = \frac{1}{ku_*} ln\left(\frac{z_{0m}}{z_c}\right) \quad [5.7]$$

According to **Hicks et al. (1987)**, this relation may be approximated such that:

$$r_b = \frac{2}{ku_*}\left(\frac{S_c}{P_r}\right)^{2/3} \quad [5.8]$$

with S_c and P_r being the turbulent Schmidt number and Prandtl number, respectively. For a gas, we have $S_c = \nu_{air}/D_i$ with ν_{air} being the kinematic viscosity of air, and D_i being the molecular diffusivity of the species. We have $\nu_{air} = \mu_{air}/\rho_{air}$ with μ_{air} ($kg.m^{-1}.s^{-1}$) being the dynamic viscosity of air.

$$\mu = \mu_0 \frac{T_0 + C}{T + C}\left(\frac{T}{T_0}\right)^{3/2} \quad [5.9]$$

with the reference values for air being T_0 = 291.15 K, C = 120 K and μ_0 = 18.27×10^{-6} Pa.s (i.e. $kg.m^{-1}.s^{-1}$).

For aerosols, r_b depends primarily on the diameter of the particle and represents the combined effect of three different resistances: Brownian diffusion, the impaction of the particles and interception **(Seinfeld and Pandis 2006)** such that:

$$r_b = \frac{1}{u_*\left(S_c^{-2/3} + 10^{-3/S_t}\right)} \quad [5.10]$$

with S_t being the Stokes number and S_c being the Schmidt number (dimensionless). For aerosols, the diffusivity in the Schmidt number will be considered as Brownian diffusion and will be estimated as per the Stokes–Einstein relationship:

$$D_i = \frac{k_B \times T \times C_c}{3 \times \pi \times \mu_{air} \times D_p} \qquad [5.11]$$

where k_B is the Boltzmann constant, $k_B = 1.38065 \times 10^{-23}$ J.K^{-1} (with 1 J = 1 kg.m^2.s^{-1}). The Stokes number is calculated as follows:

$$S_t = \frac{v_s \times u_*^2}{g \times \nu_{air}} \qquad [5.12]$$

with v_s being the sedimentation velocity. This velocity, also called gravitational velocity, is calculated only in the case of particles. It represents the effect of gravity and will gradually act upon a suspended particle. The particle will fall more and more rapidly until it achieves a threshold velocity, the *terminal settling velocity*, denoted by v_s. It has been shown that this threshold velocity is reached very quickly: for particles with a diameter of between 0.05 and 50 μm, the characteristic time ranged from 4×10^{-8} to 7.7×10^{-3}s (**Seinfeld and Pandis 2006**). Relative to the model's integration time ($\approx$1 h), it can be considered that v_s is reached instantaneously and is expressed as:

$$v_s = \frac{1}{18} \frac{D_p^2 \rho_p g C_c}{\mu} \qquad [5.13]$$

where ρ_p is the density of the particle; D_p is its median diameter (MMD); C_c is a slip correction factor; g = 9.81 m.s^{-1}, the acceleration due to gravity; and the kinematic viscosity of air, $\mu = 1.8 \times 10^{-5}$ kg m^{-1} s^{-1}. The coefficient C_c is applied to remedy a shortcoming related to Stokes' formulation, making it possible to calculate the drag force of air on a particle. The original Stokes formulation was:

$$F_{drag} = \frac{1}{8} \pi C_D A_p \rho u_\infty^2 \qquad [5.14]$$

The parameters used are as follows: A_p is the aspect ratio of the particle compared to the flux. Obviously, in the case of spherical particles this term is constant (that does not need to be determined here, however). u_∞ is the flux velocity (or velocity of the particle in a fixed milieu) and μ is the kinematic viscosity of air. However, when the diameter of the particle D_p is of the same order of magnitude as the mean free path of the particles (λ), the drag force is lower than that expressed through the Stokes equation [5.14]. This can thus be rewritten as:

$$F_{drag} = \frac{3\pi\mu u_{\infty} D_p}{C_c} \quad [5.15]$$

with:

$$C_c = 1 + \frac{2\lambda}{D_p}\left[1.257 + 0.4exp\left(-\frac{1.1D_p}{2\lambda}\right)\right] \quad [5.16]$$

The difficulty in estimating λ led to experiments that made it possible to directly tabulate C_c as a function of D_p (for a spherical particle and T = 298 K). This function has an asymptote at $C_c = 1$ for $D_p \rightarrow +\infty$. It can be noted that this corrective formulation does not have any effect when the particle size is greater than 10μm. On the contrary, for a particle with a diameter smaller than 0.1μm, the impact of this correction reduces F_{drag} by a factor of 3. We must therefore estimate λ. The calculation of the mean free path of the particles (in the first approximation) can also be considered to be equal to the mean free path of air. Most models use a mean value of $\lambda = 6.6 \times 10^{-6}$ cm. The value of λ can also be estimated more precisely using the following relationship, which makes it possible to take into account the dependence on pressure (very sensitive) and temperature (not very sensitive):

$$\lambda = \frac{2\mu_{\text{air}}}{p\sqrt{\frac{8M_{\text{air}}}{\pi RT}}} \quad [5.17]$$

where M_{air} is the molecular mass of air (28.8), T is the temperature (in Kelvin), p is the pressure in Pa, μ_{air} is the viscosity of air ($\mu_{\text{air}} = 1.7210^{-4}gcm^{-1}s^{-1}$). The calculation of λ shows that its value decreases with pressure, going from 4.10^{-6} cm for a pressure of 2.10^{4} Pa to 10^{-6} cm for a pressure of 10^{5} Pa.

5.1.3. *In-canopy resistance* r_c

In-canopy resistance is only present for gaseous species, as it represents gas absorption by vegetation. It depends on many factors related to exchanges between plants and the layer of air just above. One of the formulations that is widely used in models is that given by **Erisman et al. (1994)** and the developments carried out in the EMEP model (**Emberson et al. 2000; Simpson et al. 2012**). The expression uses different resistances and we also require chemical parameters **(Erisman et al. 1994)**. This formulation from **Erisman et al. (1994)** is as follows:

$$r_c = \left[\frac{1}{r_{sto} + r_m} + \frac{1}{r_{inc} + r_{soil}} + \frac{1}{r_{ext}}\right]^{-1} \quad [5.18]$$

where r_{sto} is the canopy stomatal resistance, r_m is the mesophyll resistance, r_{soil} is the soil resistance, r_{inc} is the aerodynamic resistance in the canopy and r_{ext} is the cuticle resistance. Depending on the type of surfaces, we may take certain shortcuts in the calculation if the grid represents liquid surfaces, $r_c = r_{wat}$, or bare soil (rocks, sand), $r_c = r_{soil}$, or snow-covered surfaces, $r_c = r_{snow}$. Based on the scheme used in **Simpson et al. (2003)**, we distinguish between bulk canopy resistance, R (and its associated conductance $G_c = 1/R$) and the resistance of the unit-leaf-area, r (and it associated conductance $g = 1/r_{sto}$). The canopy conductance is expressed as:

$$G_c = LAI \times g_{sto} + G_{ns} \quad [5.19]$$

with LAI being the leaf area index in m^2.m^{-2}, g_{sto} being stomatal conductance and G_{ns} being non-stomatal conductance. The leaf area index requires databases of maps, and for extended periods, ideally monthly. These data will represent the variability of the surface associated with the resistance.

Dry deposition depends on the state of the surface and on the turbulence in the surface layer. This process combines small-scale processes from the molecular scale to the scale of a leaf for gases. There is still very little information on many processes and these are therefore poorly represented in models. For instance, the influence of the canopy on turbulence close to the ground, or the capacity of soils to re-emit a gas that has just been deposited. A few models exist but they are very complex and dedicated on a small spatial scale. Therefore, they are not yet used in regional pollution models. Considerable developments and the implementation of many specific measurements must still be carried out to further the study of dry deposition.

5.2. Wet deposition

Open-access publication: *Zhang et al. (2019)*.

Wet deposition generally describes all processes that lead to the elimination of gases and aerosols from air through water-based processes: precipitation, cloud scavenging, snowfall, mists, etc. **(Fowler et al. 2009)**. We consider three types of sinks (Figure 5.3): (i) (*in-cloud scavenging*), which describes the capacity of a gas or an aerosol to be absorbed into or become a condensation kernel within a cloud before precipitation; (ii) (*below-cloud scavenging*), which describes a gas or a particle carried below the cloud by precipitating drops of water; and (iii) scavenging by convective updrafts **(Seinfeld and Pandis 2006; Brasseur and Jacob 2017)**. In each case, this deposition is considered to be a terminal sink. Unlike emissions whose total mass is unknown and free, deposition can only act on concentrations in the air: it is limited to what is already present in the air. In modeling, the relevance of calculating wet deposition will depend on (i) the manner in which we model the impact of a drop of water on a molecule or a particle and (ii) upstream, the capacity of the model to accurately represent clouds and precipitations.

This process is a sink term, which is expressed as:

$$\frac{\partial c}{\partial t} = -\lambda c \qquad [5.20]$$

with λ being the scavenging coefficient in s^{-1} **(Tsyro 2002)**. A priori, this is a reversible sink **(Seinfeld and Pandis 2006)**. However, in practice, through simplification, most models consider this to be a definitive sink for chemical species.

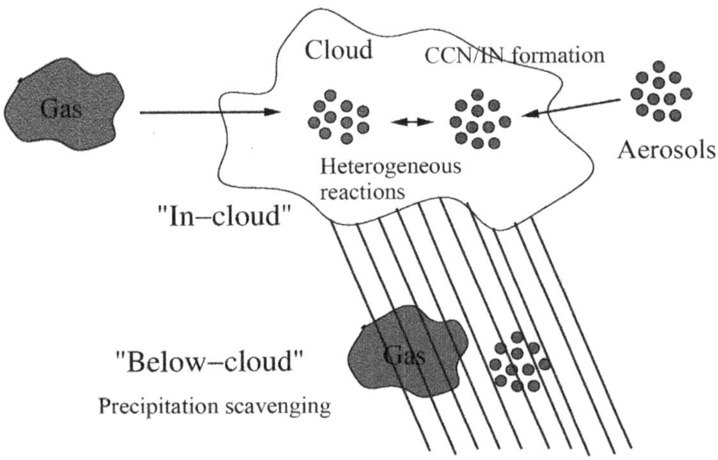

Figure 5.3. *General principle underlying the scheme used to calculate the wet deposition of gases and particles. For a color version of this figure, see: http://www.iste.co.uk/menut/modeling.zip*

5.2.1. *In convective columns*

Scavenging in convective columns is linked to gas species and particles being dragged along by updrafts during deep convection events **(Mari et al. 2000; Yang**

et al. 2015). Depending on the altitude, the type of scavenging must take into account the various phases possible between warm clouds (liquid water) at the base and cold clouds at the top (ice).

5.2.2. *In the cloud*

In-cloud scavenging or *rainout* was formalized, notably by **Giorgi (1986)**, for a global model for general circulation. This scheme remains a standard that is widely used in models. Using simulations of mineral aerosols, **Guelle et al. (1998)** were able to show the importance of accounting for these processes. The gases and aerosols may be scavenged in clouds through two different processes: for gaseous species, we consider only soluble species as they quickly dissolve in cloudwater. Aerosols may become condensation nuclei (CCN or IN) or may be scavenged by drops that are already formed by impaction. Wet deposition efficiency is highly dependent on cloud liquid water content (CLWC).

In-cloud deposition is an especially important process for soluble species like HNO_3, NH_3 and SO_2. These species may be transported with the cloud in interstitial air, or be scavenged by cloud droplets. For HNO_3, it is an irreversible sink. **Seinfeld and Pandis (2006)** showed that the scavenging rate is of the order of $\lambda = 0.2s^{-1}$.

5.2.3. *Below the cloud*

There are many parameterizations for below-cloud deposition (**Duhanyan and Roustan 2011**). The general principle remains the same: we have to model the probability that a drop of precipitating water will impact a gas or aerosol molecule as it falls. For aerosols, this probability changes with the size of the aerosol. To illustrate this kind of calculation, we present below the parameterization given by **Loosmore and Cederwall (2004)**, which is simply and commonly used in models for aerosols. The calculation of the scavenging coefficient (λ) is written as:

$$\lambda(D_p) = \frac{\pi}{4} D_r^2 V(D_r) E(D_r, D_p) N(D_r) \qquad [5.21]$$

with D_r being the representative diameter of the drops of water and D_p being the diameter of the particle. Knowing that the rate of precipitation (J in mm.h^{-1}) can also be expressed as a function of the diameter and the number of drops, it is possible to reformulate λ as a function of J and the size of particles (D_p) such that:

$$\lambda(D_p) = \frac{3}{2} \frac{E(D_r, D_p)}{D_r} \times J \qquad [5.22]$$

with $E(D_r, D_p)$ being the efficiency of the collision of particles of size D_p by a particle of size D_r:

$$J = \frac{\pi}{6} D_r^3 V_t(D_r) N(D_r) \quad [5.23]$$

Since we consider that a particle reaches its terminal velocity (V_t) almost instantaneously, we assimilate the velocity used in equation [5.23] to V_t. Determining D_r is dependent on J, according to **Loosmore and Cederwall (2004)**, such that:

$$D_r = 0.97 J^{0.158} \quad [5.24]$$

leading to a variability in the size of the drops between 1 and 2 mm. Combining equations [5.23] and [5.24] makes it possible to rewrite the expression for $\lambda(d_p)$ in a simpler form, so that:

$$\lambda(d_p) = 1.546 E(D_r, d_p) J^{0.842} \quad [5.25]$$

Following **Loosmore and Cederwall (2004)**, we use the following empirical relationship:

$$V_t = 4854 D_r exp(-195 D_r) \quad [5.26]$$

The collision efficiency was parameterized by **Slinn (1983)** such that:

$$E = \frac{4}{ReSc}\left[1 + 0.4Re^{1/2}Sc^{1/3} + 0.16Re^{1/2}Sc^{1/2}\right] + 4\phi\left[\frac{1}{\omega} + (1 + 2Re^{1/2})\phi\right] + \left(\frac{St - S^*}{St - S^* + \frac{2}{3}}\right)^{3/2} \times \left(\frac{\rho_p}{\rho_w}\right)^{1/2} \quad [5.27]$$

with the following parameters:

– $Re = D_r V_t \rho_{air} / 2\mu_{air}$ the Reynolds number of the drop of water;

– $Sc = \mu_{air} / \rho_{air} D_r$ the Schmidt number of the particles;

– $St = 2\tau(V_t - u_t)/D_r$, the Stokes number of the particles, with τ being a characteristic relaxation time;

– $\phi = d_p / D_r$ the ratio of the diameters;

– $\omega = \mu_{water} / \mu_{air}$ the ratio of the viscosities of water and the particle being considered;

and with:

$$S^* = \frac{1.2 + \frac{1}{12} ln(1 + Re)}{1 + ln(1 + Re)} \qquad [5.28]$$

One of the terms that remains to be determined is τ (in s). This is the characteristic relaxation time of the particle being studied. This time is what we considered previously in the calculation of v_t, making it possible to consider that the particle would reach its terminal velocity very quickly. According to **Seinfeld and Pandis (2006)**, we have:

$$\tau = \frac{m_p C_c}{3\pi\mu D_p} \qquad [5.29]$$

with m_p being the mass of the particle. This gives values of 4.0×10^{-8}s (for $D_p = 0.05\ \mu$ m) to 7.7×10^{-3}s (for $D_p = 50\ \mu$ m).

The scavenging is taken into account grid cell-by-grid cell for each atmospheric column. This means that in each box we consider whether there is precipitation, then whether the aerosols contained in this box are removed. We do not explicitly calculate the fact that an aerosol will descend into a lower box, until it reaches the surface in a terminal sink: this process is very rapid and establishing the scavenging for each box independently results in the same equation.

The wet deposition problem remains an open question and the research community focusing on this is very small compared to research groups for other processes. Not only are there many uncertainties regarding the mechanism itself, in and below the cloud, but these uncertainties also have a bearing on our ability to accurately model the cloud itself, using the meteorological model. In addition to this uncertainty, we have uncertainty regarding the plume of pollutants that may be scavenged: Are the gas and aerosols at the right altitude? Is the cloud at the right altitude? Has the precipitation been modeled at the right time and right place? Combining small-scale meteorology and chemistry, this process is currently one of the most uncertain and least-studied processes in the modeling system.

6
Chemistry-Transport Modeling

6.1. Principles underlying chemistry-transport models

Chemistry-transport models (CTMs) seek to represent the spatial-temporal evolution of chemical species in the atmosphere. The term "chemistry transport" brings together two large processes: firstly, the **chemistry** of a species, that is, its evolution, which depends on other species via chemical reactions; secondly, **transport** is the evolution of the species from one grid cell to the next, whether through advection, turbulent diffusion or deposition (wet or dry). The evolution of a chemical concentration (c) within the atmosphere obeys the law of conservation of mass and is expressed as:

$$\frac{\partial c}{\partial t} + \nabla.(\rho c V) = P - L \qquad [6.1]$$

where the differential represents the temporal variation in the chemical species (c). $\nabla.(cV)$ is the transport of this species by the wind V, its spatial variation; P and L schematically represent the source term (P = Production) and sink term (L = Loss). Whether directly or indirectly, this dynamic will play a role in the process being studied. To resolve this balance equation, we must know the meteorological variables to calculate the transport, the emission flux of the chemical species to obtain a source term and we must have a chemical mechanism to calculate the reactions between species and calculate the depositions. The calculation of the different parts of this equation is concretely shown in Figure 6.1. We find the "meteorological" part, mentioned earlier, as well as the necessary databases (roughness, surface type, initial and boundary conditions). We also find the emission fluxes of these chemical species, with very different origins, and dependent on meteorological conditions. Finally, the

Modeling of Regional Atmospheric Pollution,
Laurent MENUT. © ISTE Ltd 2024.

CTM will yield a result in the form of a 4D field (x,y,z,t) with different chemical concentrations. The equation can thus be written as **(Seinfeld and Pandis 2006)**:

$$\begin{aligned}
&\frac{\partial c}{\partial t} + u_x \frac{\partial c}{\partial x} + u_y \frac{\partial c}{\partial y} + u_z \frac{\partial c}{\partial z} && \text{Time increment} \\
& && \text{+transport} \\
&= \frac{\partial}{\partial x}\left(K_x \frac{\partial c}{\partial x}\right) + \frac{\partial}{\partial y}\left(K_y \frac{\partial c}{\partial y}\right) + \frac{\partial}{\partial z}\left(K_z \frac{\partial c}{\partial z}\right) && \text{Turbulent diffusion} \\
&+ \chi(c) && \text{Chemistry} \\
&+ E(c) && \text{Emissions} \\
&- D(c) && \text{Deposition}
\end{aligned} \quad [6.2]$$

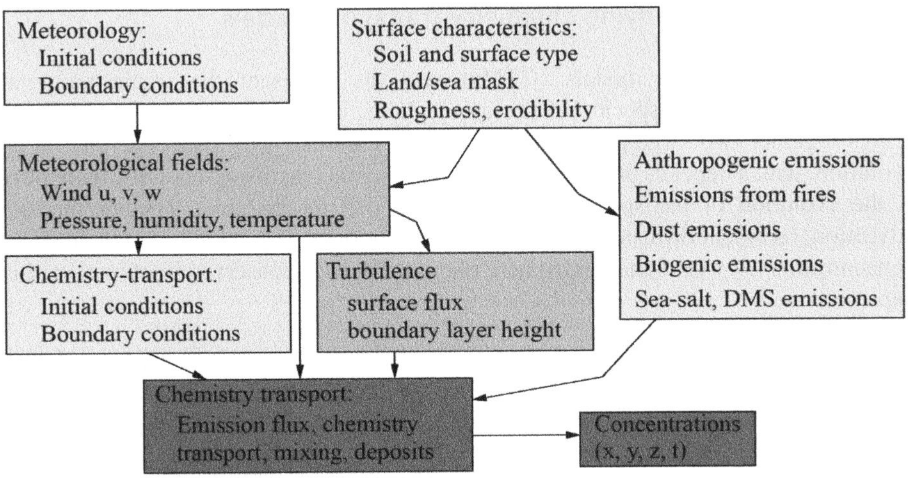

Figure 6.1. *Scheme for the working principle in an "offline" CTM. The yellow boxes correspond to the forcings in the model (initial conditions, boundary conditions, meteorological conditions and emissions), the blue boxes correspond to the meteorology and the green boxes correspond to the chemistry-transport model itself. For a color version of this figure, see: http://www.iste.co.uk/menut/modeling.zip*

There are many CTMs that are routinely used for analysis or forecasting. Table 6.1 presents a few models that are currently being used. For the rest of the chapter, the explanations and examples given will be based largely on the CHIMERE model. However, the majority of what we present is common to all of these models. Figure 6.1 presents a specific type of CTM: an *offline* model. The dynamic is calculated **before** the chemistry is calculated. The chemical model is then "fed" with the result file from the meteorological model. For some years now, *online* models have also been available: in this case, the dynamic is calculated **at the same time** as the chemistry (see section 6.6). For each time-step in the model, we integrate a complex system of equations that include both thermodynamic variables and chemical variables *(Baklanov et al. 2014)*. The case of the *online* model can be seen as a refined version of the offline model, which also conserves its particular features.

Name	Website
CAMX	http://www.camx.com
CHIMERE	https://www.lmd.polytechnique.fr/chimere
CMAQ	https://www.cmascenter.org/cmaq
GEOS-Chem	http://acmg.seas.harvard.edu/geos
LOTOS-EUROS	http://www.lotos-euros.nl
MOZART	https://www2.acom.ucar.edu/gcm/mozart
WRF-chem	https://ruc.noaa.gov/wrf/wrf-chem

Table 6.1. *A few chemical-transport models currently being used*

At present, the offline mode is most frequently used for forecasting, but this is not the case for analysis, where most models now are online models. There is a historical reason to why offline models remain dominant: meteorology modeling has nothing to do with the modeling of atmospheric chemistry. These two scientific communities have been separate for a long time. There are also other reasons: separating the tasks during calculation makes it possible to have greater flexibility for tests and validation (why should the weather be recalculated each time, when we are focusing on chemical mechanisms?). This also holds true for the modeling of the whole set of processes, if we only wish to vary the chemical-transport process. Online modeling is more difficult to implement, but is the more rigorous approach. Reality itself is online as there exists feedback between chemistry and meteorology. Regardless of whether a model is online or offline, the operations must always be carried out in the following order: (1) calculating the dynamic, that is the principal thermodynamic quantities of flux. During this calculation, we also calculate turbulent fluxes. If the model is offline, then we must conserve these quantities to calculate the chemistry next. (2) Constructing emissions: The emissions depend on the meteorology (temperature and wind) and must therefore be calculated after this is established. (3) Calculating concentrations: the CTM will take into account all contributions from sources and sinks.

For each kind of contribution, Table 6.2 summarizes the influence of the dynamic and whether we have product and loss terms to be calculated. We first see that all the processes that must be calculated depend simultaneously on the main variables of atmospheric flux (u, v, w, T, q, P) and on the associated turbulent parameters (friction velocity u_*, the boundary layer height $\overline{h}$, clouds). It can also be noted that for the product and loss terms, all processes have both terms, except emissions, which are only a source, and deposition, which is only a sink.

Variable	Taking into account the dynamic		Product	Loss
	Direct u,v,w,T,q	Indirect $u_*, \overline{h}$		
Transport	Transport	Mixing/diffusion	P_{adv}	L_{adv}
Chemistry	Reaction rates	Radiation	P_{chim}	L_{chim}
Emissions	Biogenic	mineral dust	P_{emis}	
Dry/wet deposition	Precipitations	Deposition velocity		L_{dep}

Table 6.2. *Synthesis of production and loss terms*

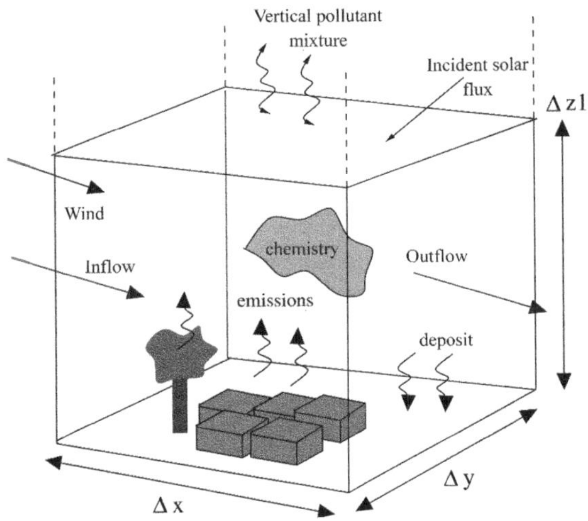

Figure 6.2. *Processes that can modify the value of a chemical concentration in a CTM. For a color version of this figure, see: http://www.iste.co.uk/menut/modeling.zip*

Figure 6.2 depicts a typical grid in a CTM and all the processes that could change the concentration value of a pollutant. The inflows are largely transport (depending on the direction of the wind) and emissions. The outflows are also transport toward the edge of the mesh and also upwards in the atmospheric column and deposits (dry or wet). In the mesh itself, the concentrations will only be modified through chemical processes.

Considering that we have calculated each term that contributes to the modification of a chemical species, it must be possible to implement a system that will integrate these quantities for all chemical species. There are two methods, predicted in Figure 6.3, to manage the system in equation [6.1]. First of all, we have the *operator-splitting* method: this is the most widely used method as it makes it possible to carry out a segmented management of all of the terms. It is less rigorous than global integration, as the result may differ based on the order in which the operations are carried out. On the other hand, it has a lower computation time than global integration as the integration over the chemical time-step is only done for chemistry. The second method *prodloss* (i.e. the global flux integration) constructs the concentration for the next time-step in an exact manner. However, this calculation takes more time because the calculation time dedicated to chemistry is often in excess of 80% of the overall calculation time of the simulation. Whether or not the other processes in the chemical time-step are integrated, it does not drastically change things. For greater rigor, it is therefore strongly recommended that we use global integration techniques for the processes. It can be seen that unlike in meteorology, CTMs often use two different time-steps: the physical time-step and the chemical time-step, as chemistry has time constants that are much smaller than in physics. While it is possible to reasonably use a time-step of 5 minutes to describe convection in the boundary layer, this is impossible when processing chemical reactions that are much quicker: the chemical time-step never exceeds 30 seconds. Thus, we must decouple the calculations (prodloss) or carry them all out with $\Delta t = 30$ s at most (*operator-splitting*). Each approach has its own advantages and drawbacks and current models use one of these two methods.

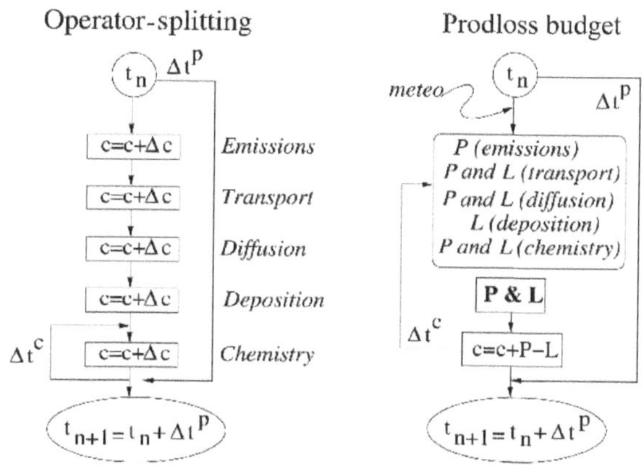

Figure 6.3. *Comparison of the principle between operator-splitting and global integration (prodloss)*

6.2. Initial conditions and boundary conditions

Whether we are carrying out a global or a regional simulation, we must have initial conditions. For a regional simulation, we must also have boundary conditions. There are different ways of supplying this information to the model being used, but there are a few principles that must be respected. The initial conditions must be prepared for a single time window, the initial time-step. We must provide a 3D field with information on the entire domain of simulation. These initial conditions will be important for yielding concentration values within the model. Indeed, even apart from pollution events, the atmosphere always has a nonzero concentration of gases and particles. Thus, at the very least, a simulation must start with values close to the climatological mean values. Boundary conditions are also particularly important for CTMs. The simulation can be forced with climatologies or simulations from the model, over a larger domain (*nesting*): the important thing is to be able to specify the import of plumes with large concentrations. Unlike in meteorological models, there is no concept of forcing within the domain during the simulation (*nudging*). This is important in meteorology as we must have a reminder of the small-scale simulations with respect to the large-scale circulations. However, in CTMs, the problem is of a different nature: the emissions bring in the required information throughout the simulation and the large domain offers no additional information with respect to the small domain beyond the edges.

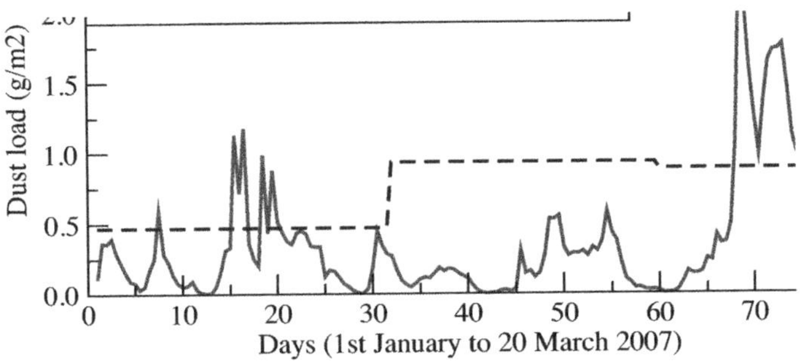

Figure 6.4. *Sensitivity of the concentrations of mineral aerosols to the frequency of the initial and boundary conditions. For a color version of this figure, see: http://www.iste.co.uk/menut/modeling.zip*

A few principles must be followed in order for the initial and boundary conditions to be optimal. First of all, we must have information on the maximum number of chemical species. At the very least, we must have information on the species being studied and with a long lifetime: ozone for pollution, CO for fires, aerosols for minerals, etc. For secondary species, the ideal is to have information at the boundaries, calculated using the same chemical mechanism: the concentrations therefore strictly represent the same chemical species. For the boundary conditions, it is recommended that we have the highest temporal frequency for the forcing. Ideally, we must have hourly windows to have a realistic variability. An example is given in Figure 6.4. Unlike ozone, the concentrations of mineral dust are sporadic across time and space.

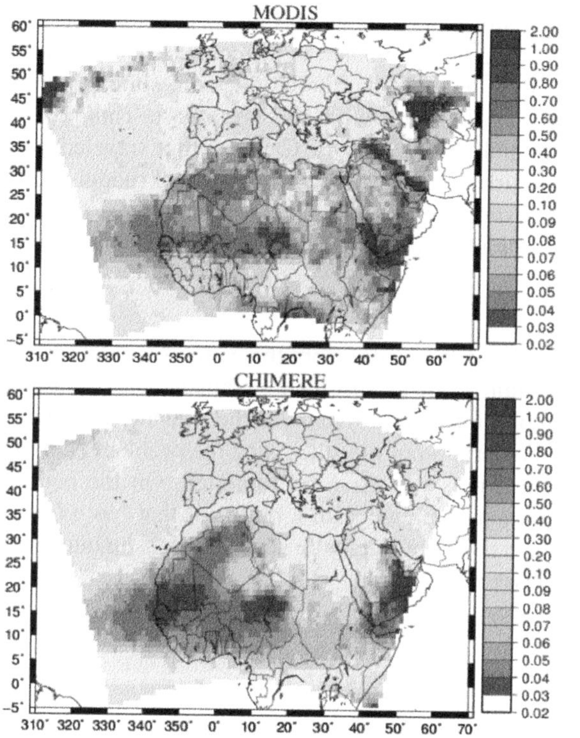

Figure 6.5. *Comparison between the measured optical thickness (by MODIS) and the modeled optical thickness (by WRF-CHIMERE) as per* Menut et al. (2015a). *For a color version of this figure, see: http://www.iste.co.uk/menut/modeling.zip*

The GOCART curves represent the monthly climatology obtained using a global model, which can force the regional CHIMERE model. The red curve represents the simulation at the same site, but directly with the regional model. It can be seen that the very large regional variability cannot be reproduced in the global model. In this case, we cannot therefore stop with a simulation forced by climatology. We must necessarily have a model on a larger scale that explicitly takes into account mineral aerosol sources. For example, to accurately simulate particles in Europe at an hourly frequency, we must carry out a simulation forced by a model with a larger domain, which calculates emissions and transport over Africa, which is the main global source of minerals. Finally, the forcing at the boundaries yield data with a vertical resolution that is close to that of the host model in order to prevent diluting information on dense plumes of gas and aerosols. Figure 6.5 depicts an example of the measurements/model comparison, performed here for optical thickness (AOD). Data from a MODIS satellite is compared to the corresponding simulation carried out with

WRF-CHIMERE *(Menut et al. 2015a)*. It can be seen that the AOD related to the mineral aerosols is accurately estimated in the center of the domain. On the contrary, at the boundaries of the domain, the model does not simulate the plumes that were observed: especially in the north-west and north-east. This relates to the transport of thick plumes, notably from fires in Canada, that traveled across the Atlantic, and whose concentrations are not known in the global model used for the boundary conditions. To better simulate this episode, we should have first carried out a global simulation of the event and used these hourly windows as the boundary conditions of the regional model.

The CTM will behave differently from the meteorological model. Here, the initial conditions are less important to specify as compared to the boundary conditions. Indeed, since the pollution is a problem "at the boundaries", with emissions being active throughout the duration of the study, and the problem being slightly nonlinear, erroneous initial conditions will soon be forgotten. On the contrary, the boundary conditions are important: a plume with a pollutant that has a long lifespan entering the simulation domain must absolutely be known and taken into account if we wish to avoid significant biases in the results.

6.3. Taking into account transport

6.3.1. *Conservation of mass*

Transport represents the advection of chemical concentrations under the effect of wind. This term is both a production term and a loss term for chemical species, grid cell-by-grid cell in the model. The transport is calculated in the form of the flux of momentum ($\rho u, \rho v, \rho w$): taking into account the density of air during transport makes it possible to ensure that the surface heterogeneities are accurately taken into account and also that the difference in the density of air in the atmospheric column is accounted for. We therefore resolve the equation for (ρc) even if we only write (c) to simplify the writing. The advection equation, called the *continuity equation*, which represents the conservation of mass, is:

$$\frac{\partial c}{\partial t} + u_x \frac{\partial c}{\partial x} + u_y \frac{\partial c}{\partial y} + u_z \frac{\partial c}{\partial z} = 0 \qquad [6.3]$$

with c being the chemical concentration that is transported, and u_x, u_y and u_z being the components of wind in the directions x, y, z. For each direction (n), it is possible to thus define the mass flux $F_n = \rho c u_n$. Since the mesh in Eulerian models is (relatively) regular, we must therefore calculate these fluxes for each facet of each grid cell, as is depicted in Figure 6.6.

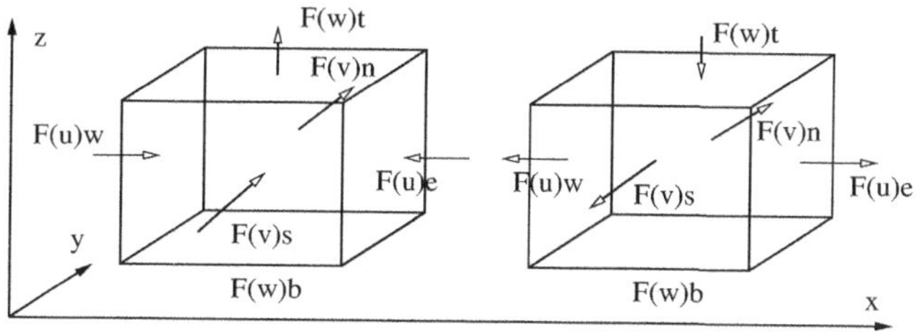

Figure 6.6. *Representation of how fluxes are processed in a grid cell in the model. The balance equations of the inflows and outflows within a grid cell make it possible to estimate the vertical flux. This principle ensures the conservation of mass for each grid cell. In this case, we consider a grid cell close to the surface: the flux from the ground is zero. (a) The majority of the fluxes bring concentrations into the grid cell. We then have a positive vertical flux $F(w)t$. (b) The majority of the fluxes cause the concentrations to flow out of the grid cell. We then have a negative vertical flux $F(w)t$*

6.3.2. *The horizontal advection flux (u,v)*

If we consider the single-dimension advection equation, we have:

$$\frac{\partial c}{\partial t} + u\frac{\partial c}{\partial x} = 0 \qquad [6.4]$$

To estimate the value of concentration in the following time-step, we must *discretize* this equation. Discretizing a differential equation to resolve it is the same as "dividing" the variations of the quantity over time and space into elementary steps. By definition, discretization is the shift from an *infinite* set of points of a function to a *finite* (discrete) set. During this "deterioration" in the quantity of information, the goal is to conserve representative points with respect to what we wish to calculate.

$$\frac{\partial c}{\partial t} = \frac{c_i^{n+1} - c_i^n}{\Delta t}; \quad \frac{\partial c}{\partial x} = \frac{c_i^n - c_{i-1}^n}{\Delta x} \qquad [6.5]$$

Here, $t^{n+1} = t^n + \Delta t$ is the time-step, and the index i represents the grid cell in space. The distance between two grid cells is Δx. Finally, the value of the concentration updated for the grid cell (i) and the following time-step t^{n+1} is written as:

$$c_i^{n+1} = c_i^n + u\frac{\Delta t}{\Delta x}\left(c_{i-1}^n - c_i^n\right) \qquad [6.6]$$

This simple scheme is explicit and a first-order scheme. It is typically used and called an *upwind* scheme **(Courant et al. 1952)**. Another formulation of the same scheme is in the implicit backward form, such that:

$$c_i^{n+1} = c_i^n + u\frac{\Delta t}{\Delta x}(c_{i-1}^{n+1} - c_i^n) \quad [6.7]$$

This writing is more robust but demands a global integration of the equation, since the concentration that we wish to estimate itself depends on the same time-step. We must therefore use Jacobi or Gauss-Seidel integration methods. In any case, a discretized equation must respect certain constraints. We must have the *convergence* of the solution to be calculated. The scheme must be *stable*, that is, the difference between the exact solution and the discretized solution must not increase with the number of integration steps over time or space. In order for the solution to be realistic, we must also respect the condition stated by Courant, Friedrich and Lewi (CFL) (1928) such that:

$$u\frac{\Delta t}{\Delta x} \leq 1 \quad [6.8]$$

that is, in a single time-step, the transported concentration must not travel a distance $u\Delta t$ (m s^{-1} s), which is larger than the size of a grid cell, Δx (m). In general, transport schemes are defined by their implicit or explicit character and their order. The explicit aspect means that we use the concentration from the time-step preceding t to calculate the concentration in the following time-step $t + \Delta t$. The implicit algorithm will use the concentration in the same time-step to re-update and is thus iterative **(Seinfeld and Pandis 2006)**. The order of a scheme is defined by the number of grid cells used for the calculation: a first-order scheme will only use the concentration in the current grid cell, a second-order scheme will use two grid cells, and so on. The higher the order, the greater the calculation time. The *upwind* scheme is known for being rapid in terms of calculation and is numerically stable, yet quite diffusive. With respect to the transport of a plume of pollutants, the concentrations will be artificially diffused horizontally in an exaggerated manner. Other schemes of a higher order have been developed to limit this numerical diffusivity. A widely used scheme is the second-order scheme by **Van Leer (1979)**. This scheme considers that the concentration in a grid cell may be represented by a linear slope (Δi) between the two interfaces. Depending on the concentration values in the grid cells $i - 1$ to $i + 1$, the value of the slope will change so that:

$$\Delta_i = sign\,(c_{i+1} - c_{i-1}) \times \min\left(\frac{c_{i+1} - c_{i-1}}{2\Delta x}, \frac{c_{i+1} - c_i}{\Delta x}, \frac{c_i - c_{i-1}}{\Delta x}\right) \quad [6.9]$$

This scheme is acknowledged to have great numerical stability, moderate calculation time and to be much less diffusive than the upwind scheme. Finally, there are also many schemes that are much more complex, but more costly in terms of calculation time, such as the third-order PPM scheme (*parabolic piecewise method*) by **Colella and Woodward (1984)**. This type of scheme is recommended for analysis simulations over short periods, but has proven to be very costly in terms of calculation time for models that carry out long simulations or operational forecasting. These schemes have been evaluated, notably for the transport of a plume of mineral aerosols in **Vuolo et al. (2009)**: it has been shown that the scheme from **Van Leer (1979)** was a good compromise between accuracy of results and calculation time.

6.3.3. *The vertical advection flux (w)*

In the case of online WRF-chem models, the mesh is the same for the dynamic and chemistry. Thus, there is no need to interpolate the wind fields onto the grids of the concentrations to be transported. On the other hand, for most offline or online models, such as CHIMERE, the meteorological and chemical meshes are different. It is tricky to use the vertical component of wind w directly interpolated, as the interpolation error may have a significant incidence on transport. It is therefore preferable to re-diagnose the value of w. This can be done as shown in Figure 6.6: w is calculated after the fluxes in the horizontal facets in order to ensure the conservation of mass. However, studies have shown that this may lead to unrealistic values of w at the top of the troposphere.

6.3.4. *Turbulent diffusivity*

In addition to the "production" and "loss" terms for concentration in a given grid cell, we can also introduce the diffusion term, such that:

$$\frac{\partial c}{\partial t} = \frac{\partial}{\partial x}\left(K_x \frac{\partial c}{\partial x}\right) + \frac{\partial}{\partial y}\left(K_y \frac{\partial c}{\partial y}\right) + \frac{\partial}{\partial z}\left(K_z \frac{\partial c}{\partial z}\right) \qquad [6.10]$$

with K_i being the diffusion coefficient in the i. It can be seen that in most CTMs the horizontal diffusion is considered to be small and already represented by transport schemes that are overly-diffusive. The term is thus neglected. We will therefore only try to calculate the turbulent vertical diffusivity (K_z). Thus, the equation to be solved is limited to:

$$\frac{\partial c}{\partial t} = \frac{\partial}{\partial z}\left(K_z \frac{\partial c}{\partial z}\right) \qquad [6.11]$$

This term, K_z, is especially difficult to parameterize. As this flux must be calculated only vertically, a question arises concerning whether explicit and implicit schemes can be used interchangeably. Let us look at a simple numerical application:

if we have a value K_z = 100 $m^2.s^{-1}$ and Δx = 5,000 m, but Δz = 50m. Knowing that we must respect the following stability criterion:

$$\gamma = K_z \frac{\Delta t}{(\Delta x)^2} < \frac{1}{4} \qquad [6.12]$$

with γ being the Fourier number, we will have a time-step that must be Δt < 17 h horizontally and Δt < 6.25 s vertically. The time-step on the vertical makes it very difficult to have realistic regional simulations. WE must therefore use an implicit scheme, such as the Cranck–Nicholson scheme (**Seinfeld and Pandis 2006**):

$$c_i^{n+1} = c_i^n + \frac{K\Delta t}{2(\Delta x)^2} \left[\left(c_{i-1}^n + c_{i-1}^{n+1}\right) - 2\left(c_i^{n+1} + c_i^n\right) + \left(c_{i+1}^n + c_i^{n+1}\right)\right] \qquad [6.13]$$

This approach is said to be *local*, since for a given grid cell we use only adjacent grid cells to diffuse a species. To update the concentration of a species at the level i, we must know the value of this concentration at the preceding time-step and at the levels $i-1$, i and $i+1$. *Non-local* approaches have been developed more recently in meteorological models and CTMs. This deviates from the hypothesis that in a highly convective boundary layer, and during a characteristic time-step of a few minutes, a concentration may be transported in the boundary layer directly from one level to another. In this case, the diffusion coefficients are estimated in a transilient matrix, which is no longer tridiagonal (the case with the local approach) but is a full matrix.

6.4. Gas-phase chemistry

The variation in the value of the concentration due to chemistry is written as:

$$\frac{\partial c}{\partial t} = P - L.c \qquad [6.14]$$

with P being the set of source terms and L being the set of sink terms related to entire group of chemical reactions. The specific formulation of P and $L.c$ is explained by the fact that emissions and chemistry may produce a quantity of c, independent of what is already present, while the loss of c can only be a fraction of what is available. If we discretize this relationship, we have:

$$c_i^{n+1} = c_i^n + \left(P_i^n - c_i^n L_i^n\right) \Delta t \qquad [6.15]$$

It can be that in the case of a *prodloss* type integration, the P and L terms bring together transformation terms related to chemistry and also to all other processes.

Unlike the processes we have studied previously, the chemical source and sink terms for a given concentration depend on the concentration of other species. This difference is important as it makes it necessary to use different resolution methods to integrate the whole chemical mechanism.

6.4.1. *Chemical reactions*

Thousands of chemical reactions act in the atmosphere on hundreds of different species. We have radical oxidation in the gaseous phase initiated by solar irradiation, oxidation in the aqueous phase within cloud droplets and heterogeneous reactions on the surface of aerosols. The transformation of a chemical species will take place through a reaction, which is the balance equation for the formation and rupture of various chemical bonds. For example, the reaction $[2A + B \rightarrow E + F]$ may be the balanced equation of intermediary reactions such that: $[A + B \rightarrow C]$ since $[A + C \rightarrow D]$ and then $[D \rightarrow E + F]$. It can be noted that C and D are successively produced and consumed, they are defined as the intermediate products of the reaction. The set of detailed reactions is the **reactional mechanism**. The challenge in accurately modeling this chemistry is therefore being able to carry out calculations with only a few reactions: the realistic transformation of the species of interest. The chemical mechanisms used in the model are reduced mechanisms: we will try to restrict it to about 10 species and reactions in order to have a system that can be computed at the scale of present-day computers. The reactions have different reaction rates and this creates priorities in the chemical kinetics. For example, for the following system:

$$\left\{ \begin{array}{lll} A + B \rightarrow C & v_1 = k_1 C_A C_B & \text{slow} \\ A + C \rightarrow D & v_2 = k_2 C_A C_C & \text{fast} \\ D \rightarrow E + F & v_3 = k_3 C_D & \text{fast} \end{array} \right. \quad [6.16]$$

The first reaction is the slowest: it will be kinetically deterministic for the complex mechanism. The resulting reaction will be:

$$2A + B \rightarrow E + F \quad [6.17]$$

with $v = v_1 \approx k_1 C_A C_B$, which leads to an overall slow rate.

When one of the intermediate reactants is consumed as quickly as the product, it can be considered that it is at an equilibrium with respect to the whole mechanism: the species is in a quasi-stationary state. The kinetics are highly dependent on temperature and for many reactions, the reaction rate is expressed as per the Arrhenius equation:

$$k = Ae\frac{-E}{RT} \quad [6.18]$$

with E being the activation energy (in kJ.mol^{-1}), A (in s^{-1}), and E and A being empirical factors. Each elementary reaction is a unique chemical event corresponding to the breaking of a chemical bond or the collision between two molecules. Table 6.3 presents the expression for the different reaction kinetics depending on the number of molecules in the system. It can be noted that the rate of a unimolecular reaction only depends on the concentration of reactant A. The rate of a bimolecular reaction depends on the product of the concentrations of reactants A and B. Trimolecular reactions are rare, since the probability of the simultaneous collision of three molecules is small. Finally, we have a system made up of different chemical species, and each of these can be involved in multiple chemical reactions with different kinetics.

Elementary reaction	Molecularity	Reaction rate	k
A → products	Unimolecular	$v = -\frac{dC_A}{dt} = kC_A$	t^{-1}
A +A→ products	Bimolecular	$v = -\frac{1}{2}\frac{dC_A}{dt} = kC_A^2$	t^{-1}C^{-1}
A +B→ products	Bimolecular	$v = -\frac{dC_A}{dt} = -\frac{dC_B}{dt} = kC_AC_B$	t^{-1}C^{-1}
A +2B→ products	Trimolecular	$v = -\frac{dC_A}{dt} = -\frac{1}{2}\frac{dC_B}{dt} = kC_AC_B^2$	t^{-1}C^{-2}

Table 6.3. *Reactions with multiple elements and corresponding kinetics*

6.4.2. *Solvers*

It is therefore necessary to integrate the system of differential equations from equation [6.14] with a solver that makes it possible to consider it to be a stiff differential system. This is a stiff system because the time constants for the various differential equations in chemistry have several orders of magnitude of difference: for example, from a few milliseconds for radicals, to a few years for methane. Solvers have been constructed to numerically resolve a system like this *(Zhang et al. 2011)*. These solvers are complex and there are many of them. They must respect a minimum number of rules: (1) the integration method must be numerically stable regardless of the integration time-step; (2) the concentrations calculated must always be positive; and (3) we must conserve the mass of the different species.

Among the most widely used solvers, we have *quasi-steady-state-approximation* (QSSA) **(Hertel et al. 1993)** and two-step process **(Verwer 1994)**. The QSSA scheme calculates the concentration in the following time-step such that:

$$c_i^{n+1} = \frac{P_i^n}{L_i^n} + \left(c_i^n - \frac{P_i^n}{L_i^n}\right) exp\left(-L_i^n \Delta t\right) \quad [6.19]$$

Depending on the lifetime of the species $\tau = 1/L$, the calculation is simplified. If $\tau < \Delta t/10$, we consider that the lifetime is short compared to the time-step and the expression is reduced to:

$$c_i^{n+1} = \frac{P_i^n}{L_i^n} \qquad [6.20]$$

If, on the contrary we have $\tau > 100\Delta t$, then the species has a long lifetime compared to the time-step and we apply the advanced Euler scheme:

$$c_i^{n+1} = c_i^n + (P_i^n - L_i^n c_i^n)\,\Delta t \qquad [6.21]$$

In the intermediate case, the complete relationship is used. The solver is quick in terms of calculation time and numerically stable, however it does not seem to perfectly conserve mass. A second possible approach is the two-step scheme:

$$c_i^{n+1} = \frac{4}{3}c_i^n - \frac{1}{3}c_i^{n-1} + \frac{2}{3}\Delta t\left(P(c_i^{n+1}) - L(c_i^{n+1})c_i^{n+1}\right) \qquad [6.22]$$

This equation can be reformulated according to **Verwer and Simpson (1995)** such that:

$$c^{n+1} = \left(I + \frac{2}{3}\Delta t L(c^{n+1})\right)^{-1}\left(\frac{4}{3}c^n - \frac{1}{3}c^{n-1} + \frac{2}{3}\Delta t P(c^{n+1})\right) \qquad [6.23]$$

The obtained system is nonlinear and implicit as it has the form $c^{n+1} = F(c^{n+1})$. It can thus be resolved using a Gauss–Seidel integration method. The advantage of the two-step process is such that the chemical production (P) and loss terms (L) are not necessarily constant during the calculation time $(t^{n+1} - t^n)$. The Gauss–Seidel method is an iterative method to solve a linear system written in the matrix form.

6.4.3. *Chemical mechanisms*

The processes presented earlier are applicable to all chemical species present in reality, that is, tens of thousands of different species. The computation resources in actual models are such that it is impossible to explicitly simulate this large number of species. And in the context of analyzing or forecasting pollution, there is very limited interest in obtaining information that is as fine or abundant. "Real" chemical species are thus grouped together into families of species that represent them: "model" species. Depending on the chemical mechanism used, this group may make it possible to have only a few tens or hundreds of species to be calculated. Certain species of interest remain true species, such as ozone or nitrogen oxides. VOCs are often grouped and there are different approaches to grouping *(Stockwell et al. 2012)*. There are three steps involved in the construction of these mechanisms: (1) collecting laboratory measurements for the largest possible number of chemical reactions; (2) constructing the most explicit mechanism possible (thousands or millions of reactions), and using these in 0D to see the functioning; and (3) based on the mechanism, choosing simplifying hypotheses to obtain a mechanism that can be integrated into a 3D CTM.

Mechanism	Model	Reference
CBM05	WRF-chem, CMAQ, C-IFS	**(Sarwar et al. 2008)**
Melchior	CHIMERE	**(Lattuati 1997)**
MOZART	ECHAM, REMO, WRF-chem	*(Emmons et al. 2010)*
RADM2	WRF-chem	**(Stockwell et al. 2012)**
RACM	COSMO, RegCM-chem, CMAQ, WRF-chem	*(Stockwell et al. 1997)*
SAPRC07	CMAQ, CHIMERE	**(Carter 2010)**

Table 6.4. *Mechanisms in gaseous chemistry used in chemistry-transport models*

Among the "explicit" mechanisms, we see the most widely used ones are the NCAR Master Mechanism **(Madronich and Calvert 1990)**, the Master Chemical Mechanism *(Saunders et al. 2003)* and the GECKO model *(Aumont et al. 2005)*. Based on these standard mechanisms, we define the lightest, as presented in Table 6.4. The best-known grouping approach is the *carbon bond* one, which groups species based on their carbon number, thereby generating model species such as alkanes, alkenes, aromatic compounds, etc.

Knowledge concerning the modeling of gaseous chemistry has now attained a level of maturity. For many years now, most CTMs use proven models. However, there are still some chemical pathways that are poorly known, studied in simulation chambers. Between these updates and the increase in computation resources, gas-phase chemistry mechanisms continue to evolve, even though the change is much slower than for aerosol chemistry. It is important to continue to work on Master Mechanisms, which alone can generate the evolution in simplified mechanisms.

6.5. Aerosols

Modeling aerosols is the most complex part of a CTM model **(Seinfeld and Pandis 2006)**. There are many processes that act on many types of aerosols (see section 1.4). We will describe below the mode of representation of aerosols and then the most widely used schemes for the transformation processes.

6.5.1. *Size distribution*

As an aerosol can have several possible sizes, a representation of the aerosol must account for this variability, as shown in Figure 1.10. Models use several possible representations: *bulk*, *modal* or *bin* distribution *(Riemer et al. 2019).*

– *Bulk* representation consists of only having a single mass concentration per aerosol. This form was used for the earliest model to save on computation time. However, as it is not very realistic it is no longer very widely used.

– A *modal* representation consists of representing the concentration of a species in the form of several log-normal distributions. For many models, there are three modes used, as they represent the most measured sizes for certain processes. These are called *fine* modes for the Aitken mode, *medium* for the accumulation mode and *coarse* for primary aerosols such as mineral dust. The advantage of this representation is that we have only three variables to represent an aerosol. The drawback is that we hypothesize that the size distribution necessarily has the same shape as the three distributions.

– *Bin* representation consists of defining class intervals for the size. For each *bin,* we use the mass median diameter to characterize the aerosol size. The advantage is that we can have any distribution and it can thus be better adapted to reality. The drawback is that there is a minimum number of bins required for the representation to be realistic. In practice, a minimum of 10 bins are required, from 0.01 to 40 μm, so that the model can accurately represent the size distribution of an aerosol.

Depending on the process to be represented, the model can also use a **number distribution** or a **mass distribution**. Most models use a mass distribution, which is required to accurately calculate emissions, transport and deposition. However, for certain chemical processes, such as nucleation, it is preferable to use a number distribution. The transition from mass to number takes place through a calculation that directly involves the diameter of aerosols and their density, however this results in numerical errors. The most recent models have schemes that explicitly calculate mass and number (thereby applying processes such as transport to both populations): these remain uncommon as they require greater computation time. However, these are developing more and are becoming more well known, given that a combined knowledge of mass and number is important for the accuracy of schemes concerning cloud microphysics *(Li et al. 2008).*

6.5.2. *Internal/external mixing*

The numerical representation of aerosols must choose between representing **internal** or **external** mixing *(Riemer et al. 2019). External mixing* is the simplest: we consider that for each size class, a given aerosol has a unique chemical composition. While this may not be the most realistic conception, it makes it possible to simplify the transport calculation, as well as the calculations for the radiative effects of aerosols:

we can use a Mie code for each aerosol, independently, and consider that the radiative impact is the sum of the impacts of all aerosols present in the mesh *(Riemer et al. 2019)*. Another representation uses *internal mixing*: for each size class, an aerosol represents a mixture of chemical compounds. Thus, for the optical properties, we must calculate a mean refractive index for every possible composition of the aerosol. This approach takes significantly more computation time. A third representation, *core-shell*, is more realistic. It uses a representation found in the atmosphere: the aerosol is composed of a core: a primary aerosol like black carbon, organic carbon or a mineral dust. A *shell* forms around this core, made up of secondary particles (sulfates, nitrates, ammonium, secondary organic aerosols (SOA), sea-salt) and water. The optical properties are therefore a mixture of the optical properties of the constituents **(Péré et al. 2010)**. In addition to the ease of calculation and the accuracy of the result, choosing this representation will also have an impact on the use of the aerosol that is calculated: *Riemer et al. (2019)* showed that internal and external mixing give different results in terms of aerosol activation for CCN (*cloud condensation nuclei*) calculation in cloud microphysical schemes.

6.5.3. *Aerosol models*

Open-access publication: *Semeniuk and Dastoor (2020)*.

The term *aerosol model* denotes the calculation of all of the processes that affect the aerosol formation cycle: nucleation, condensation and coagulation **(Zhang et al. 1999)**. These are complete models, often developed to be usable in 0D and integrated into the chemistry-transport codes. The aerosols will be processed depending on their nature. We can identify three families of species:

– chemically inert aerosols (or considered to be chemically inert): minerals, black carbon and sea-salt;

– inorganic aerosols: sulfate, nitrate and ammonium;

– organic aerosols.

Unlike gaseous species (for which we model tens of model species), we will model much fewer species for aerosols. As each aerosol has a size distribution to be discretized, the calculation cost for modeling aerosols can quickly become prohibitive. For certain models, like CHIMERE, the modeled aerosols are presented in Table 6.5 **(Bessagnet et al. 2004)**. In others, certain inorganic species are not necessarily taken into account, for instance, the IFS model from ECMWF *(Bozzo et al. 2020)*. The aerosol model is only applied to chemically active species like inorganic and organic compounds. For other species, such as minerals, we only process their emissions, transport and deposition in the CTM.

Figure 6.7 shows an example of a simulation that makes it possible to calculate aerosol concentrations based on their composition. For two different sites, in Corsica and metropolitan France, we distinguish the plume of mineral aerosols at altitude. Other high concentrations of aerosols are located closer to the source where they (or their precursors) were emitted.

Model species	Origin	Description	Density ρ_p	Primary /secondary
PPM	Anthropogenic	Primary material	1.50	Primary
DUST	Mineral	Mineral	2.65	Primary
EC	Anthropogenic	Black carbon (BC)	1.50	Primary
POM	Anthropogenic	Organic compound (OC)	1.50	Primary
SALT	Biogenic	Sea salts	2.10	Primary
SOA	Bio/anth	Secondary organic aerosols	1.50	Secondary
SO_4	Anthropogenic	Equiv. sulfate	1.84	Secondary
NO_3	Anthropogenic	Equiv. nitrate	1.70	Secondary
NH_4	Anthropogenic	Equiv. ammonium	1.70	Secondary
WATER	–	Water	1.00	Primary

Table 6.5. *Properties of modeled aerosols. The density (ρ_p) is expressed as the value $\times 10^3$ kg m^{-3}*

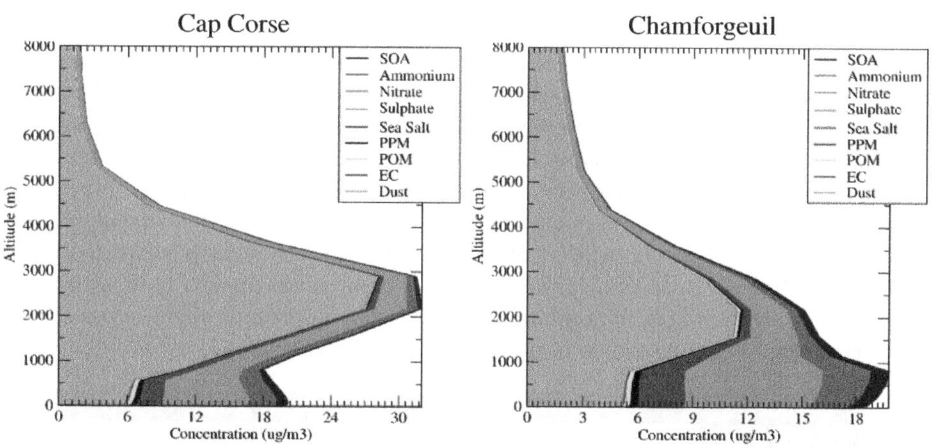

Figure 6.7. *Example the modeling of aerosol concentrations based on their composition. Simulation carried out using the WRF-CHIMERE; figure adopted from Menut et al. (2015a). For a color version of this figure, see: http://www.iste.co.uk/menut/modeling.zip*

For nucleation and coagulation processes, we usually use standard schemes in the models. **Nucleation** represents the conversion of the gas phase to particle phase. This primarily depends on the saturation vapor pressure of the molecules and the scheme by **Kulmala et al. (1998)** is very widely used. The difficulty with representing this process is that it represents the formation of ultra-fine particles, which are very difficult to measure. For a given species, the higher the saturation vapor pressure (vis-à-vis atmospheric pressure), the more volatile the species will be. This evolution is nonlinear and is quantified by the Clausius–Clapeyron relationship. The nucleation occurs as a function of the Gibbs cycle and cannot take place in an over-saturation regime. This process remains poorly known (and thus poorly modeled), especially with respect to its initiation. The saturation ratio is defined as $S = P_a/P_a^s$, with P_a being the partial pressure of the compound and P_a^s being the saturation vapor pressure of compound A, in equilibrium with its liquid phase, at temperature T. For S < 1, we will have an undersaturation regime. If the radius of the drop increases, ΔG increases: spontaneous nucleation will not be possible. For S > 1, we will have an oversaturation regime. In this case, the nucleation will evolve as a function of $r_{critique}$. **Coagulation** is more frequent than nucleation for the formation of SOA. The aerosols aggregate between themselves during thermal agitation. This phenomenon is better understood today. Coagulation occurs through **adsorption** or **absorption**, with the latter process being dominant in the system. For the coagulation process, we speak of a gas/particle equilibrium. Absorption occurs according to different laws, depending on whether the particle is hydrophilic or hydrophobic **(Pandis et al. 1993)**. For hydrophilic particles, and thus in the aqueous phase, we will use Henry's law, and for hydrophobic particles, and thus the particular phase, we will use Raoult's law. In modeling, these constants must be defined for each model aerosol species. We must therefore adapt the reaction constants to the specific mechanism used. **Condensation** is a process whereby a gaseous species will condense on the surface of an aerosol. This species may then be released in gaseous form when the cloud evaporates. The smaller the aerosol, the more rapidly this process takes place. We try to calculate the thermodynamic equilibrium on the surface of the aerosol. This equilibrium is generally calculated independently for organic and inorganic compounds. This makes it easier to distinguish between organic compounds (mostly hydrophobic) and inorganic compounds (mostly hydrophilic), as this process involves only hydrophilic. However, this has been brought into question because certain SOAs have also been considered to be hydrophilic.

Model	Reference	Used in
SORGAM	*(Schell et al. 2001)*	WRF-chem
UHAERO	*(Amundson et al. 2006)*	CMAQ
ISORROPIA 2	*(Fountoukis and Nenes 2007)*	WRF-chem, Geos-Chem, Chimere
MOSAIC	*(Zaveri et al. 2008)*	WRF-chem
EQSAM	*(Metzger et al. 2012)*	Camx

Table 6.6. *Models for calculating inorganic aerosols*

For **inorganic species**, semi-volatile species, the ISORROPIA model in **Nenes et al. (1998)** is often used, and has now been updated to version 2 *(Fountoukis and Nenes 2007)*. Other models also exist, as compared in **Zhang et al. (1999)**; **Amundson et al. (2006)** and *Semeniuk and Dastoor (2020)*. These models take into account the gas and aerosol concentrations, as well as the relative humidity and temperature. Aerosols will increase in size and it is necessary to calculate their wet diameter. This process causes aerosol diameters to change from nanometric to micrometric. Inorganic aerosols, sulfate (SO_4^{2-}), nitrate (NO_3^-) and ammonium (NH_4^+), are in liquid form and will be formed through this process from sulfur dioxide (SO_2), nitric acid (HNO_3), and ammoniac (NH_3), respectively. For **organic species**, SOA modeling has been observing great progress *(Semeniuk and Dastoor 2020)*. The complete chain of calculations is as follows: (i) emissions of gaseous precursors (biogenic, anthropogenic, fires); (ii) gaseous chemistry; (iii) nucleation by transition from gas to particles; (iv) aerosol chemistry, including in aqueous phase; and (v) physical chemistry of clouds. One of the earliest studies on the modeling of SOAs is quite recent, with *Schell et al. (2001)* and the SORGAM model. It can be noted that this modeling was first carried out on a global scale and then, more recently, on a regional scale **(Kanakidou et al. 2005)**: aerosols may be formed in the free troposphere by low temperatures, and may then have an effect on the radiative balance and cloud formation. These aerosols are formed from primary (POA) organic aerosols (VOCs), the majority of which are biogenic in origin, such as benzene, toluene, terpenes and isoprene *(Pun and Seigneur 2007)*. It must be noted that because of its small size, isoprene was not initially considered a possible precursor to the SOAs, although it is one of the most abundant *(Li et al. 2013)*. The precursor gases are first oxidized (from OH, O_3 or NO_3) to yield semi-volatile organic compounds (SVOC) **(Bessagnet et al. 2009)**. SOA modeling is sensitive to the different model parameters: (i) the quality of biogenic emissions, as these are the main precursors. In addition, we also occasionally have emissions from fires, which are important contributors of OC. (ii) The quality of the meteorological forcing data, especially the temperature close to the surface, as the gas-particle conversion depends greatly on this (through chemical kinetics). (iii) The difficulty in modeling nucleation remains a weakness in these calculations. A few current models or calculating SOA are summarized in Table 6.7.

Model	Reference	Used in
UHAERO-organics	*(Amundson et al. 2007)*	0D
SOAP	*(Couvidat and Sartelet 2015)*	POLYPHEMUS, CHIMERE
SOAS	*(Pye et al. 2017)*	CMAQ5
UNIPAR	*(Beardsley and Jang 2016)*	0D
BAT	*(Gorkowski et al. 2019)*	0D

Table 6.7. *A few models for the calculations of secondary organic aerosols*

However, most of these models have underestimated the SOAs calculated in CTMs. This may be due to an underestimation of the biogenic emissions of the precursors. However, it was identified that this was also due to the representation of the ageing of semi-volatile gaseous species before they were converted to aerosols **(Donahue et al. 2006; Robinson et al. 2007)**. To represent this ageing process, *volatility basis set* (VBS) schemes were added to SOA calculations. These made it possible to track the evolution of gaseous species over the course of their oxidation and to account for the evolution in their volatility. To do this, a part of the POA emissions (generally considered to be chemically inert and nonvolatile *(Shrivastava et al. 2011)*) were divided into semi-volatile species (SVOCs) and species with intermediate volatility (IVOC).

These VBS schemes are commonly used in models such as WRF-chem *(Shrivastava et al. 2011)*, EMEP *(Bergström et al. 2012)*, EURAD *(Li et al. 2013)*, CHIMERE *(Zhang et al. 2015)*, *(Cholakian et al. 2019)* and CMAQ *(Woody et al. 2016)*. Recently, functionalization and fragmentation were taken into account by adding these to the VBS schemes, making them more realistic: *Cholakian et al. (2019)* used both these add-ons in the CHIMERE model. Functionalization considers the POA independently, depending on their oxidation kinetics, k_{OH} and is therefore divided into categories of alkenes, alkanes and aromatic compounds. Fragmentation is taken into account based on the oxidation cycle of species.

At present, modeling the lifecycle of aerosols is one of the biggest fields of research related to pollution. There are many models of increasing complexity. The description of all of the processes becomes complex, thereby multiplying the numerical instabilities and errors in models. Current numerical approaches, such as VBS or thermodynamic models, are not yet as mature as those used for parameterizing surface turbulence or gaseous chemistry, for example.

6.6. Coupling meteorology and chemistry

Open-access publication: *Zhang (2008), Baklanov et al. (2014).*

6.6.1. *Coupled models*

Atmospheric pollution models work on both physics and chemistry. It would be entirely logical to think that each of these will influence the other **(Grell and Baklanov 2011)**. However, for a long time, modeling only considered the impact of the physics on the chemistry and not the other way. The feedback of chemistry on physics is quite a recent form of modeling and is called *online modeling*. The principle is illustrated in Figure 6.8 and can be compared with Figure 6.1.

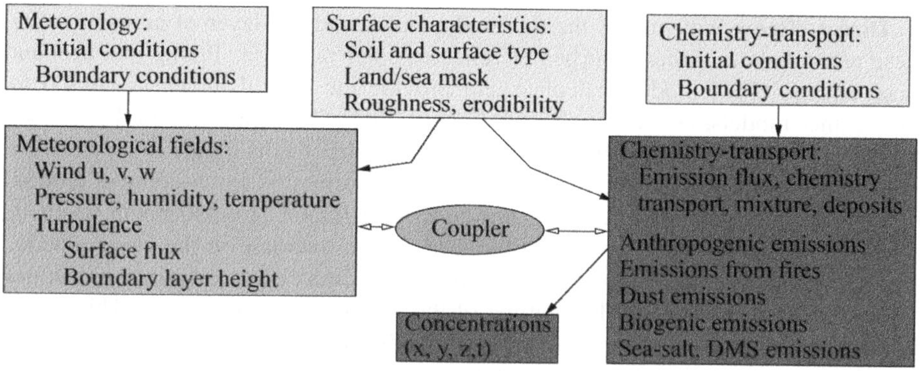

Figure 6.8. *Scheme for the working principle within an online CTM. The yellow boxes correspond to the model forcings, the blue model corresponds to meteorology and the green box corresponds to chemistry-transport. For a color version of this figure, see: http://www.iste.co.uk/menut/modeling.zip*

There are two methods for accounting for these effects in the model architecture. (i) All-in-one models: chemistry is one of the processes in the meteorological model. This means that the calculations are carried out over the same grid with time-steps that regularly overlap between physics and chemistry. This is the case with the WRF-chem model *(Chapman et al. 2009; Berg et al. 2015)*. (ii) Models with a coupler: in this case, the meteorological model is developed independent of the CTM. A third model will help them converse. The models can then also be used offline as they are not necessarily in the same grid but must have coinciding time-steps so information can be exchanged. This is the case, for instance, with models like the WRF-CMAQ *(Wong et al. 2012)* and WRF-CHIMERE, *(Briant et al. 2017; Tuccella et al. 2019; Menut et al. 2021)*.

When *online*, the two models (meteorological and chemistry-transport) carry out calculations simultaneously and exchange information all through the temporal integration. For information to move from meteorology toward chemistry, we return to the principle underlying offline models, for which the main flux variables are supplied: wind, temperature, humidity, pressure and possibly secondary or diagnostic variables, such as temperature at 2 m, wind at 10 m, turbulent fluxes, convective precipitations, stratiforms, etc. The information transfer from chemistry toward meteorology concerns gases and aerosols. The gases will act on the rate of photolysis, while aerosols will impact the radiative balance as well as cloud formation. *Online* models will calculate the possible interactions between aerosols, clouds and radiation *(Zhang 2008; Baklanov et al. 2014)*. To do this, we define two types of interactions: **direct effects** and **indirect effects (Haywood and Boucher 2000)**.

Direct effects will impact the thermodynamics under a layer of aerosols, after these aerosols have either absorbed or reflected solar radiation. Taking into account these effects, CTM models calculate and provide the optical thickness (AOD) to the weather models, as well as the *single scattering albedo* (SSA), and the aerosol asymmetry factor. **Indirect effects** will represent how aerosols influence the creation of CCN or *ice nuclei* (IN) and therefore impact the formation and lifetime of clouds. For an equivalent quantity of water, an increase in the CCN number will result in droplets that are finer and greater in number, and an increase in the cloud albedo, through what is called the Twomey effect (or first indirect effect). An increase in the CCN load will also modify the lifetime of the clouds and precipitation. This is the Albrecht effect (or second indirect effect) **(Andreae and Rosenfeld 2008)**. There have been many studies over the past few years to quantity the impact of direct and indirect effects. Table 6.8 summarizes the most recent publications describing couplings between models.

Model	References
COSMO-art	*(Vogel et al. 2009)*
WRF-chem	*(Chapman et al. 2009; Berg et al. 2015)*
WRF-CMAQ	*(Wong et al. 2012; Yu et al. 2014)*
ICON-ART	*(Rieger et al. 2015)*
Enviro-HIRLAM	*(Baklanov et al. 2017)*
BRAMS	*(Freitas et al. 2017)*
COSMO-MUSCAT	*(Dipu et al. 2017)*
WRF-CHIMERE	*(Menut et al. 2021)*

Table 6.8. *Articles describing the coupling between models to calculate direct and indirect effects*

6.6.2. *Direct effects*

The broad steps to model direct effects concern the calculation of the radiative balance. The radiative fluxes are calculated based on the temperature, humidity, liquid water/ice content and cloud fraction. We also consider the gas and aerosol concentrations that could have a radiative impact. We must therefore specify their concentrations as well as their optical properties. The forcings results are very sensitive to these choices, especially depending on whether the aerosol is described with internal or external mixing *(Péré et al. 2010)*. Among the applications on the direct effects, we mainly make note of studies on mineral aerosols, plumes of biomass fires and, to a lesser extent, anthropogenic emissions. Among the previous studies, **Pérez et al. (2006)** show, for the Mediterranean region, the increase in accuracy from taking into account the radiative impact of minerals. *Zhao et al. (2010)* studied the radiative impact of mineral aerosols in the short-wave and demonstrated how sensitive the results were to the choice of the aerosol size distribution, being able to generate a difference of up to 0.5 K for surface results. The same type of study was carried out by **Woodage and Woodward (2014)**, who showed that this choice had an impact on the simulation of the African monsoon. **Harikishan et al. (2015)** quantified the same impact on the vertical using LIDAR data and showed that the impact of mineral aerosols is more pronounced in cold clouds than in warm clouds. In Asia, **Han et al. (2013)** showed the radiative impact that mineral aerosols had on meteorology, and also showed that the simulation using an online model was closer to observations than the offline simulation. Taking into account the impact of mineral aerosols increases the stability of the lower atmosphere and therefore reduces wind velocity, which reduces mineral aerosol emissions, and thus reduces their concentration and then their transport over a large distance. *Baró et al. (2017)* studied the direct effect on temperature: they demonstrated the gain in the simulation from accounting for this effect, by applying an example over Europe. Similarly, *Briant et al. (2017)* showed an impact of the order of 1 K over West Africa. **Lou et al. (2017)** quantified the impact of the direct effect of mineral aerosols on a long simulation of 150 years, from the beginning of the industrial era. They showed a weakening in wind due to mineral aerosol emissions. *Gordon et al. (2018)* studied the direct effect of biomass fires at a high spatial resolution (4 km) and over the South Atlantic. They showed that the aerosol plume increased the CCN number, reduced the height of the boundary layer and there was general cooling over the region.

6.6.3. *Indirect effects*

Indirect effects are processed in the cloud microphysics meteorological module, a parameterization developed to simulate the lifetime of clouds, the drag at the cloud top, deep convection, precipitation and evaporation **(Thompson and Eidhammer 2014)**. A knowledge of aerosol concentrations is required for information on the CCNs **(Abdul-Razzak and Ghan 2002)** and IN **(DeMott et al. 2010)**. We must then specify both the mass and number for each aerosol size class. Chemistry in convective clouds is also important, but is not yet well described in models *(Grell and Freitas 2014)*. Studies on indirect effects are usually denoted by *aerosol radiation cloud interactions* (ARCI). One of the earliest studies was carried out by **Andreae and Rosenfeld (2008)** who presented the broad principles underlying this. *Bangert et al. (2011)* studied the indirect effects using the COSMO-art model over Europe. They showed that accounting for the relief made it possible to refine the velocities in the updrafts to enhance the calculation of the aerosol number concentrations. With these developments, they showed that it is possible to establish a link between fine-scale models (*cloud resolving models*) and global models. Their approach would then be used in other regional models. **Forkel et al. (2012)** used the WRF-chem model to establish a link between the indirect effects and impact on air quality. They showed the increase in precipitation over the Atlantic. **Lim et al. (2013)** studied the impact that this effect had on the monsoon by developing interactions in the coupled WF-CAM5 models. They showed that there was an improvement in the simulation of precipitations and the radiative balance with respect to observations. *Yu et al. (2014)* implemented indirect effects in the WRF-CMAQ coupling and applied this over the United States, comparing the results with satellite data. They showed a gain in the modeling of the PM measurements on the surface. *Nickovic et al. (2016)* studied the capacity of minerals to become glaciogenic kernels (IN). Using the WRF-chem model, *Archer-Nicholls et al. (2016)* studied the indirect effects related to biomass fires in Brazil. They showed that this generated a net cooling on the surface, as the radiative fluxes were absorbed by the aerosols. Since the aerosols from the fire had an effect on the CCN, there was an increase in the number of cloud droplets. *Zhou et al. (2016)* studied one of the most sensitive points in the calculation of indirect effects: aerosol activation. They developed ARCI interactions by applying this to the case of fog. They showed that there was a net gain in their modeling of precipitation; **Penner et al. (2018)** developed a new parameterization to better represent the indirect effects of anthropogenic aerosols (aircraft emissions) on the formation of cirrus clouds. *Weger et al. (2018)* used the COSMO-MUSCAT model to quantify the indirect effect of mineral dust on cloud formation in Europe and showed that this effect was responsible for cirrus formation. They also showed that the results were very sensitive to the choice of parameterization of glaciogenic kernels from mineral dust.

6.6.4. *Uncertainties and future prospects*

The development of calculations of direct and indirect effects at a regional level is a very current research subject. However, there are many uncertainties and thus improvements that must be made.

6.6.4.1. *The impact of the frequency of meteorological forcing*

Coupling has also become possible thanks to the recent increase in computation capacities. It became possible to exchange information between meteorology and chemistry at the frequency of a few minutes. Offline models used for analysis or forecasting, and to economize on resources, have at least an hourly meteorological forcing frequency. The modeling results thus benefit from the gain in information, by taking into account direct and indirect effects. However, it is not only this: changing meteorological forcing frequency to a sub-hourly frequency also has a significant impact *(Briant et al. 2017)*. For example, for temperature, in a convective period, an accurate description of its variation will strongly impact photolysis, while models with hourly forcing would interpolate the values between 2 hours. Similarly, for precipitation, refining the time series will drastically change the result. For all meteorological parameters with high sub-hourly variation, taking this into account has been shown to have a large impact on the simulated concentrations.

6.6.4.2. *More precision, more variability and therefore more errors*

While using an online model instead of an offline model is more realistic, there is no guarantee of obtaining better results, especially in operational forecasting. This

is because adding a link between the physics and chemistry increases the dialogue between these processes and therefore increases the possibility that an error will be propagated through the model. Thus, an error related to emissions will generate an error in the concentrations. If we add the online model, this error will impact cloud physics and therefore the radiative balance, including surface fluxes and transport, and this will in turn affect the chemistry.

Furthermore, the set of parameterizations has been developed, and thus often calibrated, using offline models. When shifting to online models, are we really in the same solution space? Normally, the physics that applies to an average (often the case in an offline model) must apply to a quantity that is much more variable. However, as many processes are threshold processes, it is probably necessary to verify the schemes that we use when we transition from offline to online.

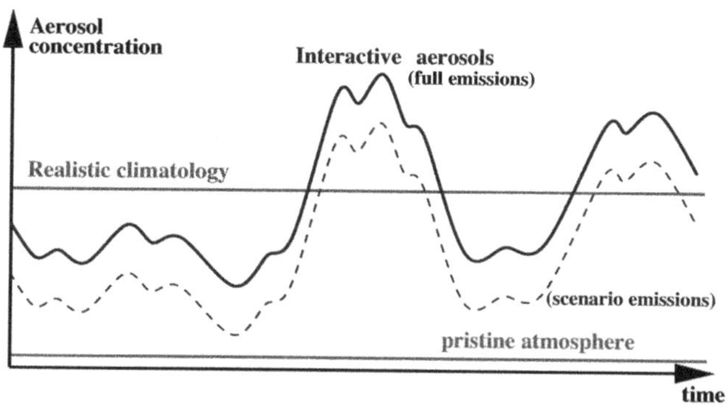

Figure 6.9. *Scheme showing the principle behind simulations with a pristine atmosphere, or climatology of aerosols, or online coupling. For a color version of this figure, see: http://www.iste.co.uk/menut/modeling.zip*

6.6.4.3. *Effects that are only partially considered*

Quantifying the gain in direct/indirect effects suffers from a definition problem: with which simulation are we comparing an online simulation? Indeed, offline meteorological models use parameterizations for which we must have information on aerosol concentrations (Figure 6.9). They thus use climatologies. Regardless of how precise they are today, these remain mean values (over days or months) and do not reflect the real situation: for example, a mean climatology cannot represent a mineral dust event or a biomass fire. To avoid this bias related to climatology, we could compare online simulations to a pristine simulation (without aerosols). However, the latter would be simply unrealistic: the comparison would have no physical sense. In practice, therefore, it is not possible to quantify the impact of an online simulation with respect to an offline simulation. However, as the online simulation is more realistic by definition, it should make it possible for simulation results to get closer to observations.

6.6.4.4. *Couplings to be developed*

The next step for couplings is to add "geophysical behavior" for the surface, including the vegetation. While global models that are used for climatic calculations already take into account these interactions, this is not necessarily the case with regional models used for atmospheric pollution. Any information that affects surface fluxes and soil states will be relevant.

Using coupled models is logical, as adopting offline models is practical but not realistic. However, all development carried out offline will probably need to be reviewed and reimagined for online modeling: putting in place multiple feedback mechanisms in a model increases its nonlinearity and therefore the possibility for errors and divergence.

7

Qualification and Optimization

This chapter presents the various approaches used to gauge how realistic a simulation is with respect to what we are trying to represent. This includes calculating statistical scores and the different representativities between measurements and observations. We then present the different methods used to evaluate the variability in the model, especially with simulations of scenarios and simulations of the whole.

7.1. Qualification of modeling results

7.1.1. *Is it possible to qualify or validate a model?*

To carry out a simulation, we must first implement a procedure that enables us to gauge its quality. The term "validation" is often wrongly used here. As a model is a simplified representation of reality, **a model cannot be validated globally**. This means that from one configuration set, the model has become universal and may yield accurate results in other very different configurations. It is possible, however, to "qualify" a model, that is, to quantify whether or not it behaves reasonably in the precise framework within which it has been defined. There are certain limitations that exist when it comes to gauging the qualities of a model: the data used to qualify the model cannot already have been used for calibrating the calculation. Further, a comparison of the model and measurements can only be carried out by keeping in mind that (i) the model contains errors: parameterizations or other hypotheses, (ii) the data contain errors related to the measuring instrument or the algorithms used to process the raw data and (iii) the two approaches never have the same spatial and/or temporal representativity.

7.1.2. *Are the results accurate?*

We must define what is an "accurate result" with respect to what we wish to study. Based on this subjective criterion, simulation results could be considered to

Laurent MENUT. © ISTE Ltd 2024.

be correct, and therefore usable for analyzing and interpreting geophysical problems. A model may be evaluated in many ways. First of all, it is possible to compare the results to known analytic solutions. The more complex the model, the less possible it becomes to use this approach. Where pollution is concerned this approach does not exist at all. It is also possible to compare the model to another model: in this case, we have an initial idea about the capacities of the model, its variability and its ability to simulate the mean values or extreme values with respect to other equivalent numerical tools. This approach is often used during exercises inter-comparing models. However, what happens when the models, often based on the same principles and the same parameterizations, yield similar but false results?

The best way to evaluate a model is still to **confront it with measurements**, which are the only reality of the process that we are attempting to represent numerically. To correctly carry out the comparisons between models and observations, the following constraints must be respected: (i) the simulated value obtained must be within the standard margin of uncertainty (the measurement); (ii) the simulated value obtained must have the same representativity as the standard; and (iii) this must be true for every point in space and for the full duration of the simulation. The measurement/model confrontation therefore brings in a "minimum" number of comparison values. This quantity remains empirical, but a few broad principles related to statistical considerations can be defined. Since the scales of space and time are related, depending on the size of the simulation domain and the simulated process, it is possible to propose bounds. For example, for a global model simulating seasonal variability, the simulations cannot be validated over a year: a system state alone is not representative. Since the global model is often used to estimate seasonal variability, the simulation cannot be considered accurate unless we have, at the least, a few years or decades of comparison. For a meso-scale model simulating an hourly field, the frequency is related to the synoptic boundary conditions: thus, we cannot validate the model over a diurnal cycle alone. The simulation is considered to be pertinent only if, at a minimum, we have several weeks or months calculated.

7.1.3. *Evaluation statistics*

There are different statistical tools to evaluate the realism of a simulation (**Wilks 2006**). These have been tested in the context of atmospheric pollution, for instance, by **Baldridge and Cox (1986)** and **Cox and Tikvart (1990)**. We use O_n to denote the nth value observed in the considered sample of size N, S_n to denote the nth simulated value and $\overline{X_N}$ for the **mean** of each of these variables, such that:

$$\overline{X_N} = \frac{1}{N}\sum_{i=1}^{N} X_i \qquad [7.1]$$

In practice, we have far fewer observations than simulated values. Thus, the dimension N is that of the measurements. The **bias** is the mean deviation between the simulated values and observed values. It indicates systematic errors:

$$bias = \frac{1}{N}\sum_{i=1}^{N}(S_i - O_i) \quad [7.2]$$

The *root mean square error* yields the mean error in absolute value:

$$RMSE = \sqrt{\frac{1}{N}\sum_{i=1}^{N}(S_i - O_i)^2} \quad [7.3]$$

Correlation makes it possible to find out whether the variability is well reproduced, independent of the amplitude:

$$cor = \frac{\frac{1}{N}\sum_{i=1}^{N}(S_i - \overline{S_N}) \times (O_i - \overline{O_N})}{\sqrt{\frac{1}{N}\sum_{i=1}^{N}(S_i - \overline{S_N})^2 \times \frac{1}{N}\sum_{i=1}^{N}(O_i - \overline{O_N})^2}} \quad [7.4]$$

It is thus possible to use these indicators directly from the values that are measured and modeled. However, we must respect a few constraints, notably the representativity of the measurements and the modeling: the two values being compared must have the same dimensions. That is:

– The *same spatial environment*: the spatial representativity of the measurement must be close to the size of the model mesh. As this is rarely the case, it is recommended that the value of the model be interpolated spatially at the site where the measurement was made.

– The *same duration*: the value of the data must be representative of the same time step as the model's time step. It is noted that measurements often have a higher frequency than the model output: the mean of the measurements will make it possible to arrive at the same representativity.

– The *same geophysical sense*: the simulated variable must represent the same thing as that which is measured. For example, we can directly compare the measured ozone and simulated ozone, but other model chemical species (or families, such as $PM_{2.5}$) cannot be directly compared to a measurement without knowing exactly what this measurement contains.

Other indicators, based on the values of the minimum and maximum diurnal concentrations can also be used **(Chang and Hanna 2004)**, as they are more indicative of what we are looking for with atmospheric pollution. Similarly, it must be considered that it is necessary to systematically use more of these indicators: a single indicator may mask problems in the simulation. Integrated numerical tools have been developed based on these indicators. For example, **Appel et al. (2011)** and **Galmarini et al. (2012)** proposed statistical models that made it possible to extract all available information in a model. **Thunis et al. (2012)** also proposed a calculation tool that was oriented more toward the calculation of thresholds intended for legislation on air quality. For example, they proposed a *target diagram* to have the bias and the RMSE on the same figure. This kind of score was used by **Simon et al. (2012)**, for instance, who summarized the performance of different chemistry-transport models using data and simulations over many years. To know whether the calculated difference between two simulations is significant, other more complex statistical tests are used. For example, the Mann–Whitney test (also called the *Wilcoxon signed-rank test*) can be used. This is a non-parametric test, that is, there is no need for the distribution of differences to be normal. We consider two simulations, $x_{1,N}$ and $x_{2,N}$, with the same dimension N. We first calculate the difference between the two:

$$d_i = x_{2,i} - x_{1,i}, \quad \text{For i} = 1, ..., \text{N} \qquad [7.5]$$

we construct a sub-set N_r by deleting the null values. The remainder is sorted in increasing order and their rank, R_i, is stored in the memory. The statistical test W is calculated:

$$W = \sum_{i=1}^{N_r} (sign(d_i) \times R_i) \qquad [7.6]$$

The z score is estimated to be $z = W/\sigma_w$ with, when $N_r \geq 20$:

$$\sigma_w = \sqrt{\frac{N_r(N_r+1)(2N_r+1)}{6}} \qquad [7.7]$$

The null hypothesis H_0 is rejected (i.e. the differences are significant and not due to chance) if $|z| > z_{critical}$. For a significance level of 0.05, we have: $z_{critical} \approx 1.645$. Figure 7.1 depicts a difference between two simulations over West Africa during the DACCIWA campaign *(Menut et al. 2019)*. It can be seen that wherever the wind value changes, there is a significant difference.

Other approaches have been proposed. For example, *Menut et al. (2017)* showed that it is possible to define a statistical indicator, comparing the model and observations, which makes it possible to quantify whether the correct scores obtained in the simulation are significant. Indeed, depending on the mode of the comparison carried out, it is possible to have correct scores but wrong reasons. For example, for ozone, if there is a small bias across a day (diurnal cycle) or over a year (seasonal cycle where the sunlight and temperature have been reasonably well calculated by the meteorological model), it is easier to model this than to accurately obtain the value of the daily peak over several consecutive days. This shows that some kinds of natural variability can artificially increase correlation. In order to avoid these "seasonal" effects, the proposed indicator compares a simulation over a period with observations for the same period but from other years. If the year that is evaluated has significantly better scores than others, this proves that the accurate result from the model is not due to the recurrence of some phenomenon from one year to the next. **Eder et al. (2014)** developed an approach using "principal components" to evaluate air quality models. This approach makes it possible to highlight the recurrent structure of the bias in the model among the apparent "noise" in the differences between the model and measurements.

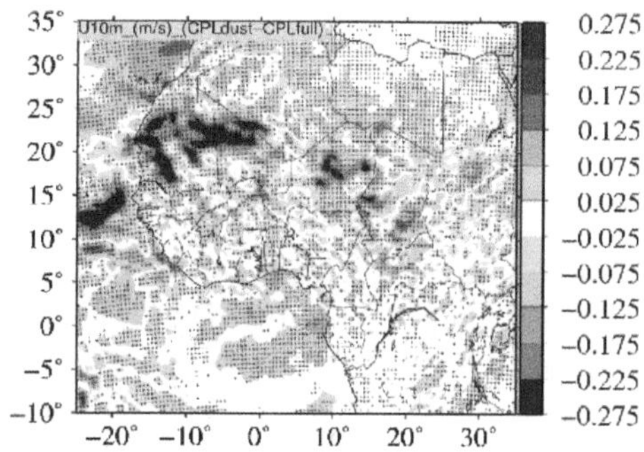

Figure 7.1. *The Wilcoxon test calculation over two simulations, with or without the direct effect of aerosols. The difference is illustrated on the wind speed at 10 m. Grid cells with a dot are those with significant differences. Figure adapted from* Menut et al. (2019). *For a color version of this figure, see: http://www.iste.co.uk/menut/modeling.zip*

7.2. Comparisons with measurements or inter-model comparisons

7.2.1. *Comparisons with routine measurements*

Comparisons with routine measurements compare modeling results with measurements from automatic stations, like the measurement of pollutants from the EMEP network, meteorological soundings, or AOD measurements from the AERONET network. These comparisons make it possible to verify the pertinence of the model over the long term, but at a high temporal frequency. Through bias and correlation calculations, it is possible to quantify the capacities of the model over a large number of different situations. Since routine measurements are usually taken by a network, in a unified electronic format, these are the data that are most often used when modeling studies are carried out **(Solazzo and Galmarini 2015)**. Recent research programs that evaluate models include the AQMEII project (Air Quality Model Evaluation International Initiative), comparing North American and European simulations **(Vautard et al. 2012; Campbell et al. 2015)**, the EURODELTA project **(Bessagnet et al. 2016)** and EMEP (European Monitoring and Evaluation Programme), a multi-year project carried out within the framework of the United Nations convention on *Long-range Transboundary Air Pollution* **(Prank et al. 2016)**. An example is depicted in Figure 7.2, which presents a comparison of ozone concentrations in a mountainous region, using high horizontal resolution (here, CHIMERE at 1 km) *(Bessagnet et al. 2020)*. For a simulation period of over a month, it can be seen that the model was able to reproduce the magnitudes and variability of ozone in Casset. In Grenoble, an urban site, the concentrations are well reproduced for certain peaks, however other days are over-estimated. This problem is typical of a turbulent mixture diagnosed too strongly in the model, limiting the titration of ozone by anthropogenic NO_x.

7.2.2. *Comparison with satellite measurements*

Figure 7.3 presents an example of the comparison between measurements by the CALIPSO satellite, with an inbuilt CALIOP LIDAR, and the CHIMERE model. For the comparison between these two tools to be possible, each of them has been transformed: (i) the CALIOP LIDAR data have been interpreted and a color code has been assigned to each type of aerosol. This differentiation is carried out from the backscattering and extinction of the LIDAR signal. (ii) The same type of classification is carried out with the model results. We then select the dominant aerosol to be represented.

7.2.3. *Comparisons with measurement campaigns*

Measurement campaigns make it possible to use research instruments complementing routine measurements. It is then possible to compare modeled variables, which are not usually found in the measurements. We will also be able to access different measurement sites, with airborne measurements, for example.

Figure 7.4 depicts an example of the studies based on the CHARMEX measurement campaigns, carried out in the summer of 2013 on the Mediterranean perimeter. The Y-axis shows the altitude of the plane in flight. The comparison between the measured ozone and modeled ozone shows that the model has an over-estimation bias close to the surface, but less so at an altitude of around 4,000 m.

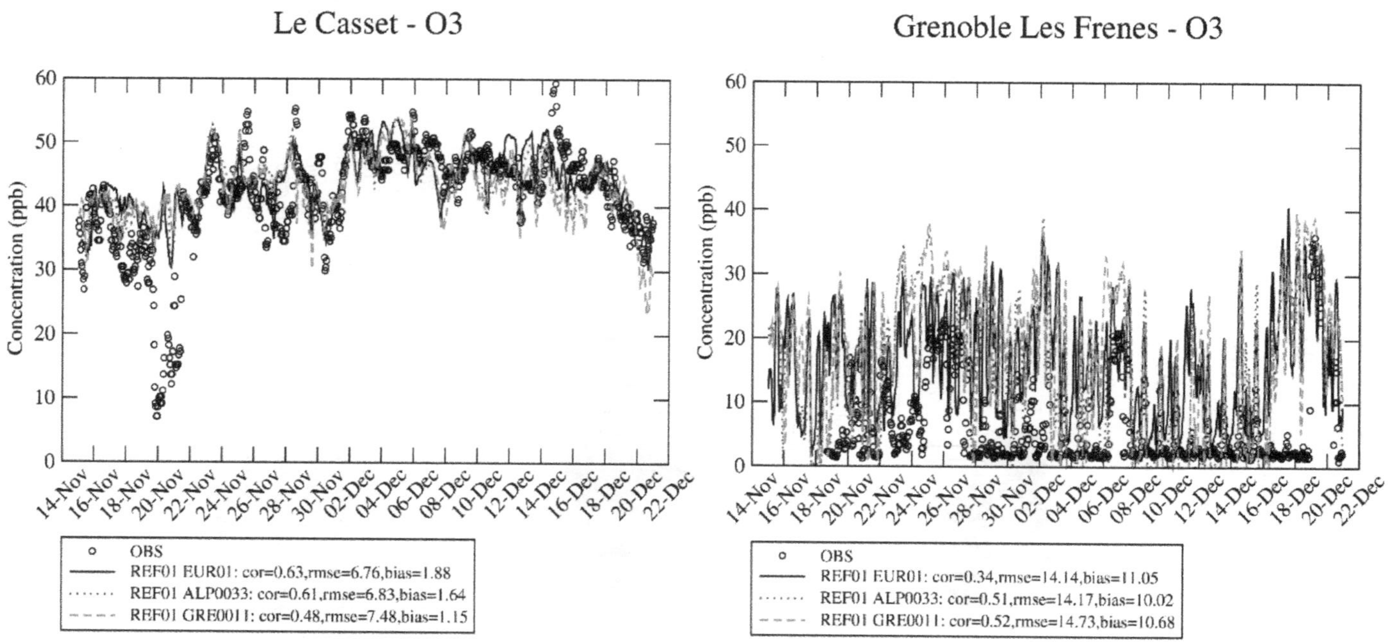

Figure 7.2. *Comparison between the measured surface ozone concentrations and simulated concentrations for two sites in montage (Casset and Grenoble). The data is taken from* Bessagnet et al. (2020). *For a color version of this figure, see: http://www.iste.co.uk/menut/modeling.zip*

The temperature is well simulated and is thus probably not the cause of the problem in the ozone modeling. The wind is relatively poorly simulated, but with a model resolution of 50 km it is normal for this highly variable parameter to not be accurately reproduced. The set of comparisons allows us to both understand the physical phenomenon as it occurred, and quantify the confidence that we may have in the model.

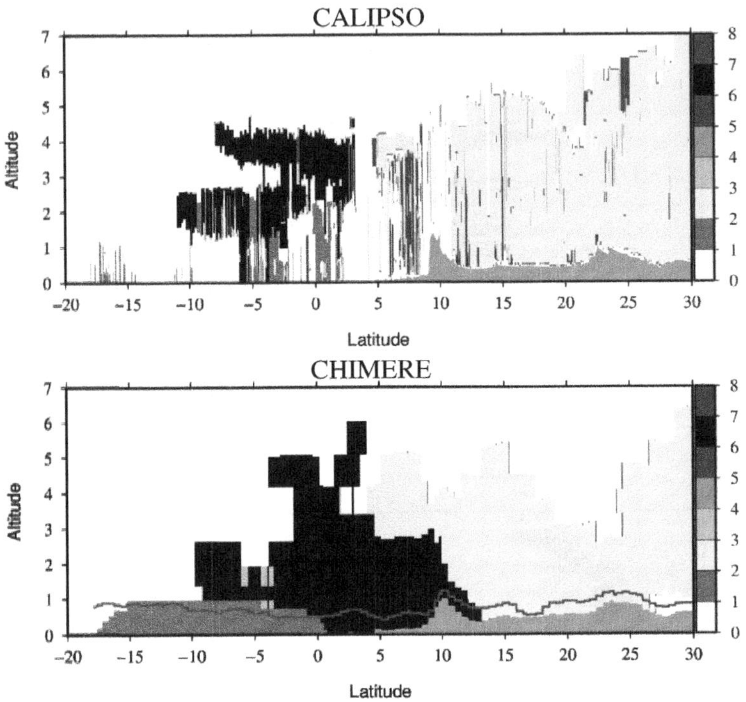

Figure 7.3. *Comparison between CALIOP measurements of aerosol types and the results from the CHIMERE model for July 27, 2014, at 01:00 UTC. The data were taken from Menut et al. (2018). The color codes correspond to (0:1) invalid, (1:2) marine aerosol in clean air, (2:3) mineral aerosol, (3:4) aerosol from continental pollution, smoke, (4:5) clean continental air, (5:6) anthropogenic dust, (6:7) smoke at altitude, (7:8) mixture of marine aerosols and dust. In the case of the model, the height of the boundary layer is added in red. For a color version of this figure, see: http://www.iste.co.uk/menut/modeling.zip*

7.2.4. *Inter-model comparisons*

Many model inter-comparison exercises are regularly carried out. If we start from the principle that there are many models to calculate chemical concentrations and these have the same goals, but not necessarily the same methods of achieving the goals, the comparisons are interesting. By comparing different models, it is possible to quantify the variability of what we are able to model. The inter-model comparisons also come under "modeling the whole set of processes" presented in section 7.6.

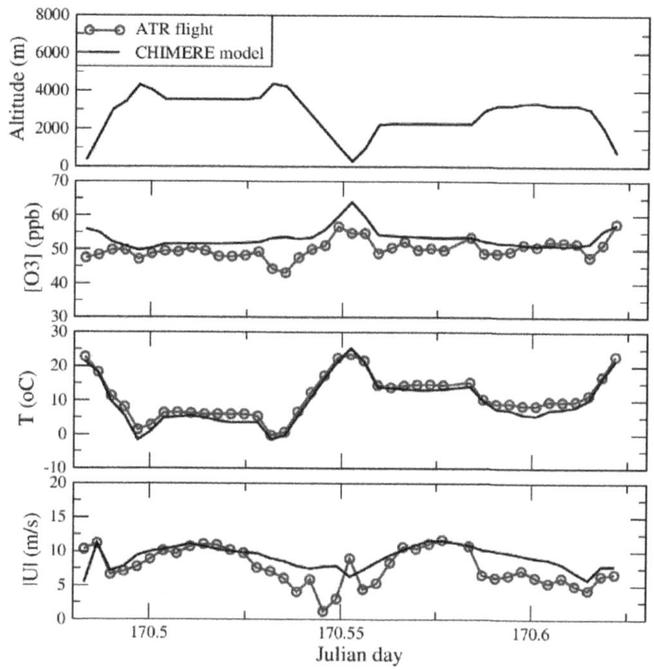

Figure 7.4. *Comparison of data from airborne measurements and the modeling equivalent using the CHIMERE model. The measurements were taken for the CHARMEX project in the summer of 2013. Adopted from from Menut et al. (2015a). For a color version of this figure, see: http://www.iste.co.uk/menut/modeling.zip*

Depending on the level of comparison between the models, we will obtain different information. Thus, inter-comparison exercises have been constructed based on very different philosophies and, consequently, yield very different results. We can list two kinds of exercises: (i) highly constrained exercises: we force the models with the same meteorology, the same emissions in order to determine the differences over a highly reduced set of processes. The aim here is to respond to a targeted question of the type, "What impact does dry deposition have on concentrations?". To respond optimally, the other parameters must be as close as possible. This is the case with the study carried out by **Vivanco et al. (2017)**. Implementing the exercise was difficult because of the need to bring together the forcings, parameterizations and resolutions. (ii) Another approach consists of leaving the models with no constraint: each does its best. The goal here is to see the diversity of models, to quantify the state of the art and to estimate what can be done optimally. This is the case with certain combined exercises used for analysis or forecasting.

To compare the model results, the best thing to do is have all outputs on a common grid, which is rarely the case. However, while the data are not optimally re-projected, the differences between the models may be due to the projection as much as due to an error in the model itself. Depending on the original grid and the target common grid, the data will need to be projected according to its resolutions. We denote the spatial or temporal resolution of the available fields by Δ_1 and use Δ_2 for the resolution of the desired fields. Depending on their respective values, various projections will be applied: if $\Delta_1 >> \Delta_2$, we can use an interpolation; this will then create information. On the contrary, if $\Delta_1 << \Delta_2$, it is possible to establish a mean value and there is thus information degradation. The principle of the mean does not pose very much of a problem and even makes it possible to estimate the induced standard-deviation (and thus the variability) from the small scale to the large scale. Figure 7.5 presents an example for both of these. This case is extremely simplified as the coarse mesh is already centered with respect to the fine mesh. On the left, the shift from the finer mesh (squares) to the coarser mesh (grayed-out circles) takes place through a simple arithmetic mean. Depending on the number of “squares” available between two “circles”, we establish the mean over four squares or 16 squares. On the right, to shift from the coarser mesh to the finer mesh, we use an interpolation: here, a simple bilinear interpolation and the segments represent the “weight” that we will allocate to each piece of information in the coarse mesh.

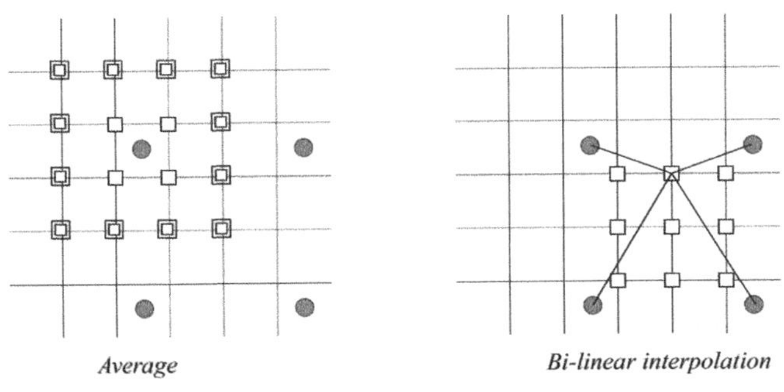

Figure 7.5. *Interpolation and mean: the problem of shifting from one mesh to another to inter-compare models. On the left: the value we are seeking is less well resolved than the input value. We can use the “nearest neighbor” technique or take the mean of all the surrounding values. On the right: the value we seek is better resolved: we use an interpolation*

Among many inter-comparison studies, we can cite the example of **Vautard et al. (2012)**, which presents a comparison of regional model meteorologies used within the framework of the AQMEII model. They showed the large variability in meteorological simulations, especially for wind in the lower layers, which is a parameter that is very important for natural emissions (dust, sea salt) and for the dispersion of emissions close to the surface. **Campbell et al. (2015)** compared indicators to evaluate how accurately models were able to reproduce chemical regimes and quantified that the models were not really able to do this. **Schaap et al. (2015)** compared the results from the models to quantify the impact of the horizontal resolution. They showed, logically, that primary anthropogenic sources are particularly sensitive to this variation in resolution.

In all cases, inter-comparing models makes it possible to obtain a more complete response than a simulation alone. It makes it possible to better define the variability in a process and its uncertainty. The inter-comparison of models without constraints is easier to carry out and is probably more advantageous than when forcings are imposed, simply because we are never able to enforce everything strictly and, consequently, never succeed in concluding that a process has been accurately isolated.

7.3. Statistical adaptation or tuning

7.3.1. *Modifications to input parameters*

Adapting or "tuning" a model consists of modifying certain input parameters of the model in order to get closer to measurements. This is not data assimilation, as there is no automatic procedure or minimization that depends on time. Instead, this is a "de-biasing" process. In simpler terms, an input parameter is modified to decrease the bias on a particular output variable. This is a very widely used approach when the aim is not to represent a process optimally, but only to optimally represent the impact it has on another process. This remains very qualitative and there is no physical-chemical knowledge that is brought into this procedure. This finds application notably in meteorology, future climatic projections and operational forecasting of pollution. For instance, *Yang et al. (2012)* modified coefficients in parameterizations of precipitation at the regional level using the WRF model. They developed a regression algorithm and modified parameters in the precipitation scheme. The results showed a gain in the wind and temperature scores. In this case, the tuning concerned uncertain and adjusted parameters in a parameterization and appeared quite legitimate. The gain may be due to the fact that the parameterization was calibrated for a specific use, a particular resolution. A different calibration was found for a different resolution. But in no case did the tuning make the model universal: it only shifted an equilibrium in one specific case toward an equilibrium in another case.

7.3.2. *Modification of results*

Statistical tuning consists of modifying the results of a simulation in order to approach measurements. In this case, the model is not modified. We only aim to correct the bias, not look for the reasons that it exists. This applies, for instance, to the output of a forecasting simulation: here, the forecast must be corrected in the same way everyday in order to increase the probabilities of having accurate scores for pollutant peaks, for example. This kind of statistical tuning is applied to the PREVAIR forecast system **(Honoré et al. 2008)**. **Demuzere and van Lipzig (2010)** and **Porter et al. (2015)** proposed simple methods to calculate the differences in concentration between the models and observed values in order to adjust the results from their model.

7.4. Scenarios

The principle underlying scenarios is to perturb an input parameter in the model to quantify the impact that this has on a desired output variable. Contrary to the tuning, we consider here that the model, in its undisturbed configuration, corresponds to the most realistic version. There are many possible applications depending on the type of modification carried out before the simulation. For example, we can estimate the impact of future weather (global climate change), future emissions (plan laws) or a process (i.e. the uncertainty in a parameterization).

7.4.1. *Meteorological scenarios*

Scenarios in meteorology may differ greatly in principle depending on whether the model is regional or global. In the case of studies on atmospheric pollution, we use regional models above all. The scenario may thus arise from (i) large-scale forcing, boundary conditions, (ii) regional forcing, like surface characteristics. In the case of studies on climate change, the scenarios are based on the radiative forcings. The study by **Frei et al. (2006)** compares several meteorological models for past and future scenarios and shows the variability in precipitation extremes. **Menut et al. (2013b)** used meteorological ranges from future scenarios and demonstrated the impact in the context of air quality: the use of a climate model as meteorological forcing changes the pollution results through the bias in wind and precipitation.

7.4.2. *Emission scenarios*

Scenarios around emissions are the most common applications. Where pollution is concerned, anthropogenic emissions remain the number one source and driving force possible.

Vautard et al. (2005) studied the 2003 heatwave in Western Europe using the CHIMERE model. They applied a reduction of 20% to the emissions, corresponding

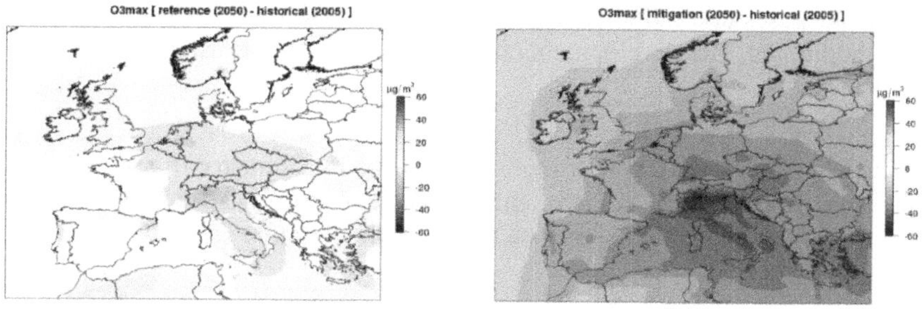

Figure 7.6. *Maps showing the differences between a standard simulation (current climate and emissions) and regional scenarios taken from CMIP5 recommendations. Figure adopted from* Colette et al. (2013). *For a color version of this figure, see: http://www.iste.co.uk/menut/modeling.zip*

to an emergency situation in Europe. They showed that the greatest impact on ozone peaks remained the decrease in NO_x. **Cofala et al. (2007)** played out emission scenarios using the RAINS model. They tested the model's response with the emission reductions recommended by 2030. They showed that the planned scenarios led to very small decreases in concentrations. **Thunis et al. (2007)** presented studies of scenarios in the context of a dedicated European project, CityDelta. By carrying out simulations with different models and different resolutions, they showed that the emissions scenarios must be carried out both at the regional and urban levels in order to be complete. **De Ridder et al. (2008)** modified the occupation of areas in their regional model in Germany. By redistributing urban populations towards the outskirts, they modified the spatial extent of the anthropogenic emissions related to urban heating and vehicular traffic (with a constant number of inhabitants). They showed that vehicular traffic increased by 17% with this scenario. This led to an increase in PM_{10} across the region, as well as an increase in ozone concentrations in the urban plume. **de Meij et al. (2009)** carried out scenarios focused on northern Italy, the Po plain being one of the most polluted regions in Western Europe. They reduced the emissions per activity sector, one by one. They showed that the most significant impact was obtained with the SNAP7 sector (road traffic): with $PM_{2.5}$ and SO_2 in winter, and with ozone in the summer. *Beekmann and Vautard (2010)* also studied the heatwave in 2003 and discussed the scenarios with reductions in emissions in terms of the variability of the chemical regime. They showed that this process must be accounted for region by region for the best possible emission reduction efficacy. *Colette et al. (2013)* sought to discriminate the strongest influence on future pollution in 2050. Using emission and meteorological scenarios, they showed that the variability in emissions remained the primary cause for an increase in pollution in the future (Figure 7.6). Another example is that of *Gidden et al. (2019)* who used various socioeconomic scenarios in CMIP6 in order to estimate the impact on emission concentrations. This was an evolution with respect to the previous exercises on studying the climate, which mainly acted on greenhouse gases and radiative balance. This time, the study focused directly on what was modifiable. The development of these scenarios will be used in many impact studies.

Emission scenarios are a robust tool, widely used in chemistry-transport models. Whether these study current cases or future climatic conditions, this tool gives the clearest view of the impact of sources on atmospheric pollution.

7.4.3. *Scenarios related to a process in the model*

Scenarios concerning parameterizations are another evaluation tool, this time to assess how realistic the model being used is. There is uncertainty regarding the accuracy of these studies, since the model is a highly simplified version of reality and thus it is important to regularly carry out exercises to quantify the uncertainty. In principle, these exercises may resemble sensitivity studies or ensemble modeling. For example, a model parameter may be modified many times to see whether or not the impact on the concentrations is linear (**Menut 2003**). Figure 7.7 depicts an example of a simulation where two parameters are disturbed at the same time. Both examples show the ozone concentration at 3.00 p.m., in one grid cell in the CHIMERE model representing Paris. Tens of simulations were carried out, with two parameters being disturbed each time: on the left, the wind speed in the surface grid and VOC emissions; on the right, the temperature at 2 m and VOC emissions. For each parameter, the disturbance is a multiplication of a factor from 0.5 to 1.5 with an increment of 0.1. This makes it possible to see, for both cases, that the variability in the ozone peak is almost linear if the parameters are disturbed up to $\pm 10\%$. This information makes it possible to make projections on the concentrations when small variations are observed in these parameters.

7.5. Sensitivity studies

Sensitivity studies have many things in common with scenario studies. However, their field of application may be much wider as we do not modify a parameter within a realistic limit (like an anthropogenic emission flux), but rather around the uncertainty related to the parameter. Thus, the values for the modifications of the parameters may be very different. For example, we can test how sensitive ozone is to temperature or cloud cover even if these variables cannot be changed in reality.

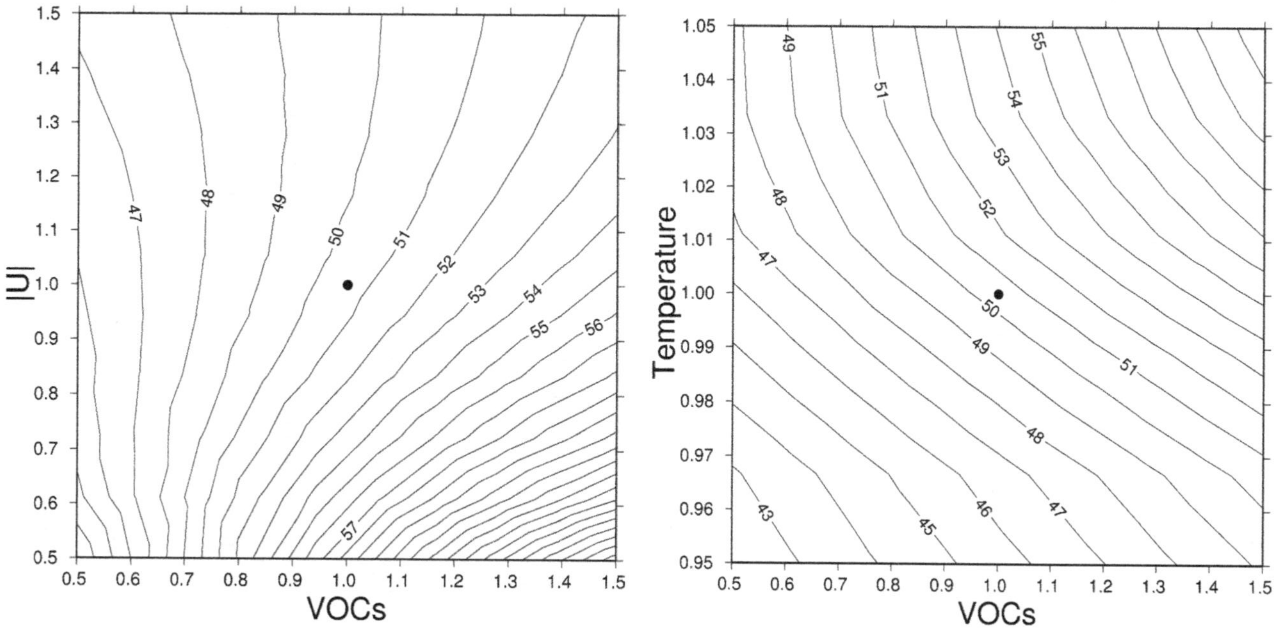

Figure 7.7. *Scenarios for the simulation of the ozone peak at 3.00 p.m.in a grid cell in a regional model representing Paris. On the left: the wind speed in the surface grid and VOC emissions. On the right: the temperature at 2 m and VOC emissions. Figure adapted from* **Menut (2003)**

7.5.1. *Sensitivity through direct modeling*

Akin to ensemble modeling or scenario modeling, sensitivity in direct modeling makes it possible to obtain the variability in one variable with respect to another. Many studies have been carried out using this methodology from the earliest use of regional modeling of urban ozone (**Al-Wali and Samson 1996; Jang et al. 1995**). Carrying out these calculations is always important as a model evolves, and we are always looking to see which parameter it is most sensitive to and whether this is realistic. Different methodologies have been developed to test the sensitivity of a model, such as the KKP system for chemical mechanisms (**Sandu et al. 2003**) or the ADIFOR system for how sensitive ozone is to the parameters of a model (**He et al. 2000**). **Hakami et al. (2003)** used the *decoupled direct method* (DDM) to calculate sensitivities: these consist of adding calculations for the gradient (the evolution of a concentration with respect to the evolution of a parameter) directly in the model, throughout its spatial-temporal integration. The method by **Hakami et al. (2004)** is the most sensitive method across several orders with high DDM. They calculated isopleths of the sensitivity of ozone to NO_x and VOCs. The sensitivity of pollutants to parameters such as meteorology continue to be studied; even recently, *Shi et al. (2020)* quantified the impact of the variability of meteorological parameters on urban concentrations of ozone and PM_{10}.

It is also useful to carry out studies on sensitivity to the architecture of the model: the horizontal and vertical resolution, the time step of temporal integration. Depending on these choices, the parameterizations implanted in the model are more or less efficient and adapted. This will then have a direct impact on the simulated concentrations. **Valari and Menut (2008)** show that there is an optimal horizontal resolution: it is beneficial to refine the mesh to better describe rapid urban chemistry. However, after a certain level of refinement (a few kilometers), the quality of results will deteriorate, as the database of forcings (including anthropogenic emissions) is not precise enough anymore, and the uncertainty around wind direction creates fine plumes, but with the wrong dispersion direction. This type of result was confirmed by **Falasca and Curci (2018)** who also tested urban schemes.

Sensitivity studies allow us to quantify the impact of one variable on another, but not only this – in the response we also aggregate the specific case of the model used to carry out the calculation, with its potential biases and variabilities. We therefore never have a purely physical result.

7.5.2. *Sensitivity through adjoint modeling*

The limitations in the previously described approaches are that a single parameter is modified to see what the effect is on one or more species. Consequently, to have a "range" of causes, we must simulate each disturbance or scenario. Adjoint modeling makes it possible to overcome this problem (**Rabier et al. 1992; Sandu et al. 2003**).

A clear description of an adjoint model is presented in *Errico (1997)*. This is the modeling used in variational assimilation (detailed in section 8.1 of this book). By definition, the **direct model** calculated concentration for each time step, depending on its value in the preceding time stop and all of the processes affecting it in the time step used in the calculation. The **adjoint model** is the gradient of the direct model: to be precise, it makes it possible to calculate the derivative of a quantity simulated with all the input parameters. This is thus a second model, which is different from the first, but which has, at every point, the same parameterizations and computation architecture from the first.

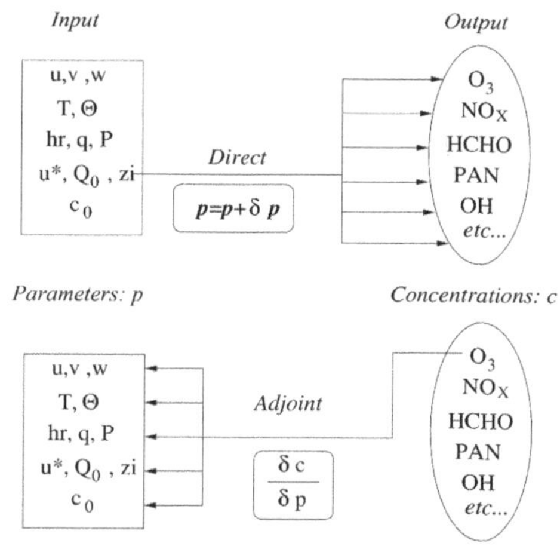

Figure 7.8. *Schematic depiction of the principle underlying the adjoint of a chemistry-transport model*

As seen in Figure 7.8, the direct model considers meteorological, emission and other parameters as input data. Via the chemistry-transport module, we calculate the chemical concentration (O_3, NO_x and others) from one time step to the next. If we wish to study a scenario, we can modify one or more input parameter. However, this modification will affect all output concentrations. On the contrary, the adjoint model will calculate the dependence of a single output species for all the input parameters. This is done in a single integration of the model (one in direct mode and one in the adjoint mode). It is thus possible to quantify, for a given species, which parameter is most important for it, and also where, when and why. The adjoint model therefore makes it possible to estimate the impact of the variation of one of the parameters for a given concentration. This control allows us to constrain the trajectory of the model in this space of parameters. The adjoint equations are obtained from the tangent linear model, which is the linearization of the direct model (nonlinear) around the input variables. The evolution of a set of variables, x, is defined as:

$$\frac{dx}{dt} = \mathbf{M}(x, p) \tag{7.8}$$

with x being the concentrations of the "state vector" of the model, p being the set of parameters of the model and **M** being the model. The discretized equation for the evolution of the model variables is $x_{n+1} = M(x_n, p)$. The adjoint model makes it possible to calculate the sensitivity of x_n with respect to p, with the latter being a vector that represents all the parameters. The sensitivity (S) of the scalar function $M(x)$ of x at a given instant t will be the gradient such that $S = \nabla_p(M(x(t)))$. At each instant t, the disturbance of x will be expressed with the following system:

$$\delta x_{n+1} = \left(\frac{\partial M}{\partial x}(x_n, p)\delta x_n\right) + \left(\frac{\partial M}{\partial p}(x_n, p)\delta p\right) \tag{7.9}$$

We will have to resolve the system given below:

$$\begin{cases} \delta^* p_n & = 0 \\ \delta^* p_{n-1} & = \delta^* p_n + \left(\frac{\partial M}{\partial p}(x_{n-1}, p)\right)^* \delta^* x_n \\ S & = \delta^* p_0 \end{cases} \tag{7.10}$$

For example, to estimate the sensitivity (S) of the ozone peak at 15.00 UTC, with r_1 being the reaction rate of the chemical mechanism, we will have the value in %/%:

$$S[O_3]_{/r_1} = \frac{\partial [O_3]_{max}}{\partial r_1} . \frac{r_1}{[O_3]_{max}} \tag{7.11}$$

Using an adjoint CHIMERE model, **Schmidt and Martin (2003)** calculated the sensitivity of ozone on the scale of Western Europe. Using the same model, **Menut (2003)** carried out the same type of calculation over the Parisian agglomeration, for ozone and for NO_2. Other studies of the same kind have been carried out with the IMAGES, POLAIR, TM4, and DRAIS models *(Nester and Panitz 2006)*, among others. In all cases, these studies were limited to gas-phase chemistry, the limitation being due to the difficulty in developing an adjoint model for aerosols, both in terms of writing and in terms of computation time, as well as the difficulty of creating adjoint versions for processes like thermodynamic equilibrium. *Henze et al. (2007)* proposed a complete adjoint, for gas and aerosols, on a global scale using the GEOS-chem model. Figure 7.9 presents a map showing the sensitivity of the simulated mass of nitrates to NH_3 emissions. For the thermodynamic module, *Capps et al. (2012)* proposed the adjoint of the ISORROPIA module, a model that is widely used in many chemical-transport models. The adjoint can also be used to calculate the sensitivity of other parameters: for example, *Karydis et al. (2012)* calculated the sensitivity of cloud water droplets to parameters for aerosol formation.

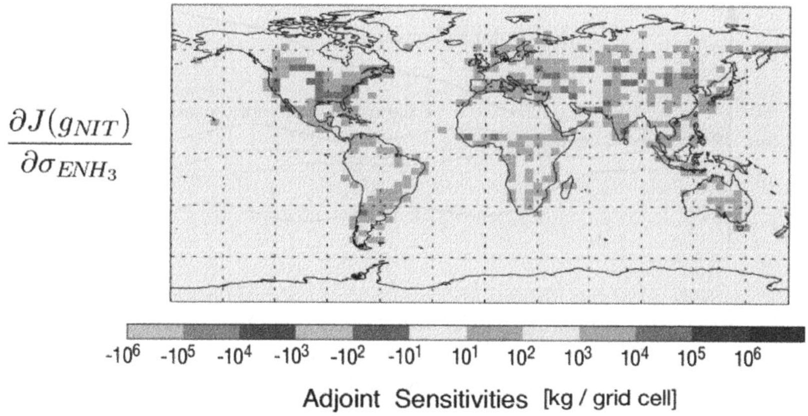

Figure 7.9. *Sensitivity of the total mass production of nitrates to NH_3 emissions. Adopted from* Henze et al. (2007). *For a color version of this figure, see: http://www.iste.co.uk/menut/modeling.zip*

Sensitivity studies remain essential for evaluating the behavior of a model. The adjoint approach is more powerful than the direct approach, but requires doubling the size of the model with a dedicated development.

7.6. Ensemble calculations

The principle behind ensemble calculation is to carry out multiple simulations of the same event and compare the results. Each simulation is called a "member" of the ensemble. It has been shown that the mean or median of the members yield statistically better scores than a single member **(Delle Monache et al. 2006)**. Ensemble modeling is implemented in two contexts: (i) studies on the uncertainties in the results and (ii) operational forecasting. In this paragraph, we will only look at uncertainty studies, as forecasting is presented in section 8.3.

As with many uses of models in geophysics, ensemble forecasting was first put in place for meteorology and then transposed to chemistry-transport. In meteorology, the formulation of the equations makes it chaotic: for very small differences in initial conditions, there may be very different results (i.e. the "trajectory" of the mode) (Figure 7.10; (**Lorenz 1963**)). In order to arrange the different possible solutions based on the uncertainty of these initial conditions, the ensemble modeling was quickly put in place for weather forecasting (**Hamill et al. 2000**). Scores, like the "ensemble skill score", have been created (**Atger 1999**). **Leutbecher and Palmer (2008)** provided a historical review of ensemble calculation methods.

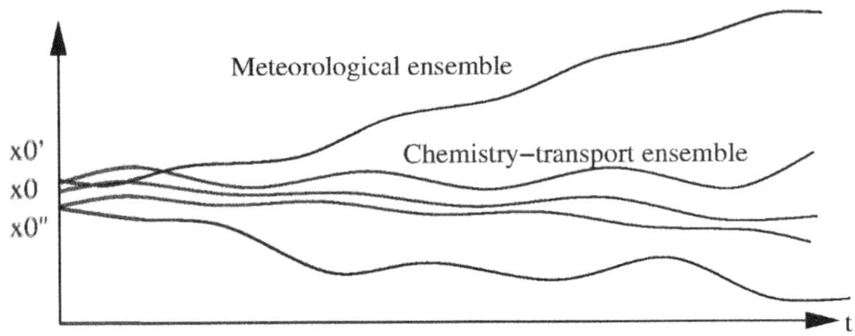

Figure 7.10. *Comparing model trajectories between meteorology and chemistry-transport. For a color version of this figure, see: http://www.iste.co.uk/menut/modeling.zip*

As forecasting results have been clearly improved with an ensemble, **Dabberdt and Miller (2000)** suggests applying this approach to the modeling of pollution. **Galmarini et al. (2004)** discusses the relevance of this kind of result for a decision-making process by showing that the analysis of the ensemble and its variability improves the reliability of the results. **Pagowski et al. (2006)** implemented one of the earliest instances of ensemble modeling for predicting ozone on the surface. Nonetheless, the very nature of the problem to be solved in chemistry-transport makes ensemble modeling less efficient than in meteorology. As Figure 7.10 illustrates, since the meteorological model is very nonlinear and primarily depends on its initial conditions, a small disturbance in these initial conditions will drastically modify the results. The chemistry-transport is slightly less nonlinear and depends on its initial conditions, but above on all its boundary conditions. Therefore, even with an initial disturbance, the model will converge toward a state of equilibrium. A gain may be brought in by creating ensembles with the weather model forcing the CTM and the CTM itself. **Zhang et al. (2007)** also showed that in the cases of ozone pollution, the wind is naturally weak. The uncertainty regarding the calculation of wind close to the surface becomes more and more sensitive and uncertain. Ensemble modeling makes it possible to give a more reliable variability for ozone predictions.

While it is clear that as with meteorology the results are better with members in the ensemble, there is still an ongoing debate over how to construct each member. There are two possible approaches: (i) the simplest approach consists of disturbing a single model in different ways; (ii) the more complex approach consists of using several models for the same case. It can also be noted that recent studies have used both approaches: **Van Loon et al. (2007)** and **Schaap et al. (2007)** used seven models over Western Europe to create ensembles for gas-phase chemistry and aerosols. They showed that the multi-model approach made it possible to access new information, the median being new information. Their results demonstrated that the median is closer to the observations than the "best" model in the ensemble. Finally, the scattering of the ensemble is a "measurement" of the uncertainty of the simulations. **Vautard et al. (2009)** also carried out simulations over Europe and used "rank diagrams" to qualify them. These were originally created for data assimilation (**Talagrand et al. 1999**). The principle for this is presented in Figure 7.11: we rank the measurement based on the results from the model, placed in increasing order in "bins". If all the results fall into the left bin, the ensemble systematically overestimates the measurement.

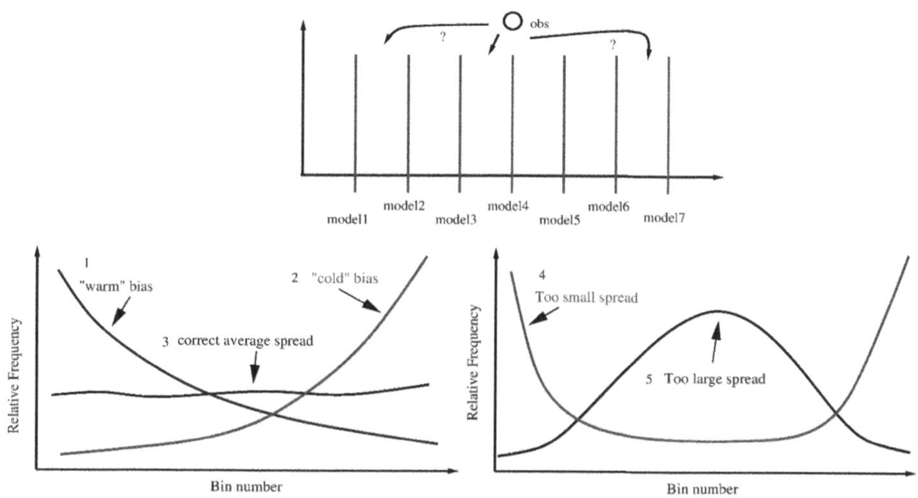

Figure 7.11. *Principle underlying rank histograms. For a color version of this figure, see: http://www.iste.co.uk/menut/modeling.zip*

On the contrary, *Garaud and Mallet (2010)* used a single model, Polyphemus, to generate members. The advantage of this approach is that a single configuration can be used to generate many members. However, the results are less chaotic than with multi-model simulation. The ensemble yields an envelope that gives the variability of the model around the reference value, but is limited to this information. *Zyryanov et al. (2012)* designed a multi-model ensemble (Figure 7.12). *Kioutsioukis and Galmarini (2014)* summarized the various studies that came earlier and presented the methodologies to use for ensemble modeling of pollution. They reviewed the different statistical metrics to optimally analyze the results of the ensemble simulations.

Figure 7.11 shows the principle underlying the classification of simulations carried out in an ensemble *(Talagrand et al. 1997)*. The principle is simple and clever. In the figure (upper row), for each simulated value (concentration of a species at a given moment and in a given grid cell), the models (1–7) can be arranged in increasing order of value. We know the corresponding observations and can therefore place them in the corresponding interval. By repeating the operation as many times as there are observations, we will have traced a rank histogram. The x-axis will show the number of bins corresponding to the models, and the y-axis shows the relative frequency of occurrence. We thus have a:

1) *Warm bias*: by analogy to a temperature simulation, a warm bias here means that all observations are lower, each time, than all the model values. We thus have an ensemble of models that overestimates the observations.

2) *Cold bias*: here we have a model that systematically underestimates the observed values.

3) *Correct average spread*: we have a flat distribution of comparisons: this is what we can ideally expect for an ensemble.

4) *Too small spread*: the observed values are either lower than or higher than the values from the models. This means that this ensemble uses models that all yield an equivalent response, creating an ensemble without enough variability.

5) *Too large spread*: contrary to the previous case, in this case the variation in the observations is smaller. The models are too variable.

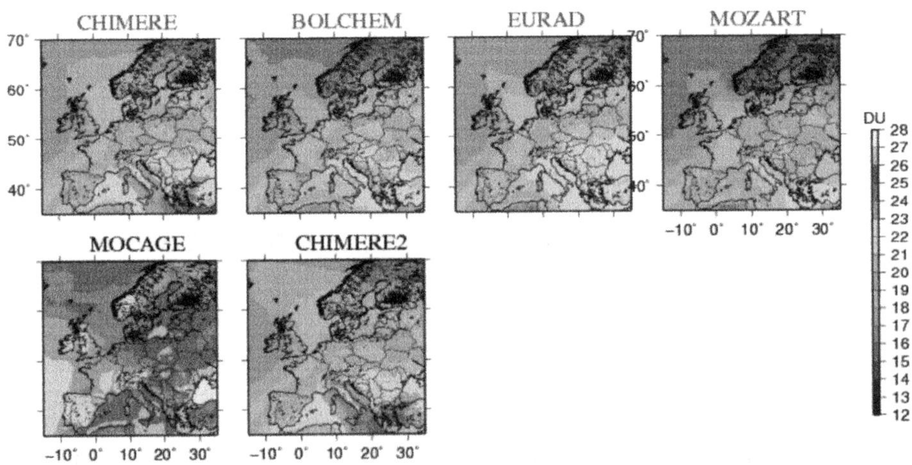

Figure 7.12. *Multi-model modeling. Integrated ozone column. Adopted from* Zyryanov et al. (2012). *For a color version of this figure, see: http://www.iste.co.uk/menut/modeling.zip*

7.7. Machine learning

The most recent developments in modeling are related to *machine learning* and artificial intelligence. As with scenarios or ensembles, this is no longer a development of the model itself, but focuses on the best way to use the model in the context of long-term simulations or projections for the future. This new field is similar to neural network methods, which have been used in geophysics for a few decades now.

Neural networks have been in existence since the 1970s and used to predict air quality. They are mathematical tools constructed for local forecasting and making use of surface observations **(Kolehmainen et al. 2001; Russo et al. 2013)**. They offer the advantage of highly rapid computations, which is an important aspect in forecasting. The drawback, however, is that they only cover existing cases: despite more and more sophisticated learning, neural networks can only reproduce situations that have already occurred. Since it largely uses surface measurements from networks, the tool is generally unable to account for long-distance transport of plumes. As it is limited to what is already known, it cannot predict inter-annual evolution in emissions (unless there is new learning every year).

More recently, we have seen the emergence of deep learning techniques based on results from deterministic models *(Bessagnet et al. 2021)*. In this case, this relates to the construction of a meta-model based on an existing chemistry-transport model. An innovative use would be to downscale the concentration fields with high precision. Based on learning from a low-resolution simulation and then a high resolution simulation, it carries out a low-resolution forecast (therefore deterministic, but rapid). The deep learning model will then make a high-resolution projection.

8

Data Assimilation, Inverse Modeling and Forecasting

This chapter describes numerical data assimilation and inverse modeling techniques. These techniques are used to improve the result or forcings in a model by hybridizing the model with measurements. We then present forecasting, which increasingly makes use of assimilation techniques these days.

8.1. Data assimilation

> Open-access publication: *Sandu and Chai (2011).*

Regardless of the method used, the goal is unique: *taking into account measurements to best describe the state of the atmosphere through the use of a model* **(Sandu and Chai 2011)**. This makes it possible to (i) better estimate a model input parameter, (ii) reduce the uncertainty in forecasts, and (iii) create "analysis maps".

During a simulation, errors accumulate in the model and the difference between the model and the measurements will always increase (Figure 8.1). Since the assimilation algorithm tries to reduce this difference, data assimilation is a minimization problem. As with other modeling approaches, data assimilation in geophysics comes from meteorology **(Talagrand and Courtier 1987)**. Computation techniques were then adapted to chemistry-transport models *(Bocquet et al. 2015).* Three broad families of data assimilation applications have been identified:

1) **Analysis**: this is used to create a database with a model of the best possible quality. For example, if we wish to study long-term trends, a simulation may be carried out over several years. Data assimilation across this entire simulation makes it possible to obtain results that are better than the so-called "first guess" **(Pierce et al. 2007)**.

2) **Inverse modeling**: the minimization between the model and the observation over the course of the simulation aims to optimize an input parameter of the model.

Modeling of Regional Atmospheric Pollution,
Laurent MENUT. © ISTE Ltd 2024.

3) **Forecasting**: with no information about the future, this approach aims to optimize the model's initial conditions at the beginning of the forecasting.

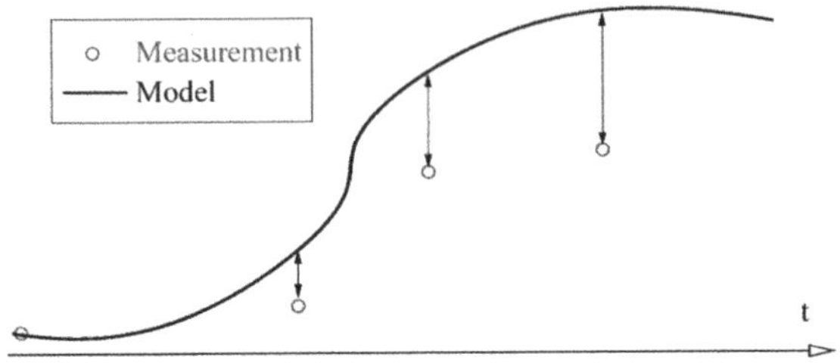

Figure 8.1. *Trajectory of the model and measurement values. The difference between the model and measurements will increase with integration time due to errors, nonlinearities and uncertainties in the model. For a color version of this figure, see: http://www.iste.co.uk/menut/modeling.zip*

There are two large algorithmic families for assimilation: sequential assimilation (SA) and variational assimilation (VA) (**Talagrand 1997**; Figure 8.2). Both types of assimilation have their own advantages and disadvantages. They can be summarized as follows: SA uses the direct model as it is and replays simulations based on measurement/model comparisons. VA will require an adjoint model. Further, SA will take into account the model error, while VA hypothesizes a perfect model.

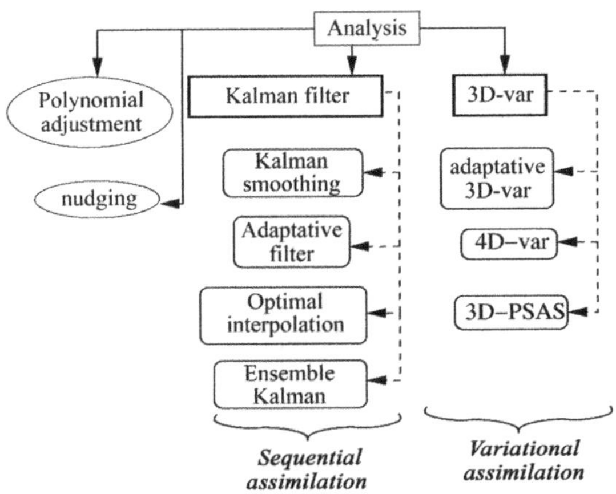

Figure 8.2. *The different algorithmic families in data assimilation*

8.1.1. *Sequential assimilation*

The principle underlying sequential assimilation is to take into account an observation at the moment when it is available a time t. The anticipated state of the model (the guess) is then corrected and the new state of the model (the analysis) is used for the calculation at the time $t+\delta t$. The state of the model is then re-updated as many times as there are available data. To get as close as possible to the observation, we use filtering techniques like the extended Kalman filter (EKF) and optimal interpolation (OI). The main advantage of the EKF is that it is (relatively) easy to implement and, most notably, that it yields the uncertainty of the analyzed result based on observation errors and model errors. On the contrary, information propagation for sequential assimilation can generally only take place from t to $t + \delta t$, that is, the new analyzed state can only depend on "past" observation. This is not a problem in the case of forecasting, but may pose a problem when we try a posteriori assimilation of data acquired over an entire period (it is then desirable for the analysis at t to be constructed with observations both prior to and following t). The general principle underlying the **Kalman filter** is explained here as an example. It offers the major advantage of providing the uncertainty in the analyzed result based on observation errors and model errors. In the next few paragraphs, we will provide the main characteristics of the linear Kalman filter. We will use the following notations:

– x^t is the real value of the desired variable;

– x^a and x^f are the corresponding analysis vector (denoted by a) and the forecast vector (denoted by f);

– P^a and P^f are the estimation error covariances for analysis and forecasting, respectively;

– M corresponds to the model, and M^T corresponds to the transpose;

– η is the model error with covariance Q;

– y^0 corresponds to the observation vector;

– H is the observation operator;

– K is the analysis gain matrix;

– x^b is the value of the guess (denoted by b, for "background") used for the analysis;

– R denotes the observation errors and I is the identity.

The notation $\overline{w}$ denotes an ensemble mean of w, that is, the mean at a given time of all possible values of its argument. Between the times t_i and t_{i+1}, the model studied is:

$$x^t(t_{i+1}) = M(t_{i+1}, t_i)x^t(t_i) + \eta(t_i) \qquad [8.1]$$

The covariance of errors η is known and is such that $\overline{\eta_{t_i}\eta_{t_i}^T} = Q(t_i)$. We also consider that the errors are unbiased, that is, $\overline{\eta} = 0$ and $\overline{\epsilon} = 0$. Finally, we consider that there is no cross-correlation between the model errors and observation errors, and therefore $\overline{\eta\epsilon} = 0$. The Kalman filter will define a set of equations that yields the optimal path for using the data and the model for the "best" estimation of the desired value. The algorithm is then written as:

1) guessed state: $x^f(t_{i+1}) = M(t_{i+1}, t_i)x^a(t_i)$;

2) covariance of guess error: $P^f(t_{i+1}) = M(t_{i+1}, t_i)P^a(t_i)M^T(t_{i+1}, t_i) + Q(t_i)$;

3) Kalman gain matrix: $K_{i+1} = P^f(t_{i+1})H_{i+1}^T \left[H_{i+1}P^f(t_{i+1})H_{i+1}^T + R_{i+1}\right]^{-1}$;

4) analyzed state: $x^a(t_{i+1}) = x^f(t_{i+1}) + K_{i+1}\left[y_{i+1}^0 - H_{i+1}x^f(t_{i+1})\right]$;

5) covariance of analyzed error: $P^a(t_{i+1}) = [I - K_{i+1}H_{i+1}]\, P^f(t_{i+1})$.

The algorithm thus consists of calculating steps (1) and (2) for each time step in the model (t_i, t_{i+1}), by inserting the evaluated corrections with relations (4) and (5) when the observations are available: we then use relation (3). The Kalman filter is often used in the form of ensembles, referred to as *EnKF*. Multiple simulations are launched with initial perturbations, and the Kalman filter is used to assimilate the data. **Constantinescu et al. (2007)** shows that this approach makes it possible to reduce the divergence that is often approached with the use of the Kalman filter on a single simulation.

8.1.2. *Variational assimilation*

The principle here is to globally adjust the trajectory of the model to all available observations in the assimilation time window. In this case, and unlike with the sequential approach, at a time t, the analysis will take into account observations preceding it and following it (**Elbern et al. 1997; Elbern and Schmidt 1999**). Using this approach, we must define a cost function J, which represents the "distance" between the observations and the modeled solution such that:

$$J = \sum_j \alpha_j \left(y_j - y_j^0\right)^2 \qquad [8.2]$$

where y_j^0 represents the observations, y_j represents the model values and α_j is the representative "weight" of the observation's accuracy. As the uniqueness of the model solution is related to the initial conditions, these conditions will act as control variables. The techniques used is to then minimize J by looking for the minimum of its derivative. Iteratively determining the minimum of this gradient of a parameter may be quite time-intensive in terms of computation time. It is therefore possible to determine this gradient more quickly, using the model's adjoint equations. Current resolution methods are called 3D-VAR and 4D-VAR (including the temporal dimension).

8.1.3. *Applications of data assimilation*

The first type of application is to enhance a simulation to spatially "fill in" the lack of measurements or create maps with fewer errors. This analysis can be made with any data (for instance, a meteorological variable or a chemical species) and anywhere in the domain of study: with surface data or satellite data. Another kind of assimilation is to "de-bias" information to better constrain a chemical species. For example, it is possible to assimilate the temperature to improve the modeling of the photochemistry of summertime ozone. It is also possible to assimilate aerosols, or the AOD (aerosol optical depth) directly, to thereby improve the photochemistry. Assimilation techniques can thus help in reconstructing maps of a space for itself or for its impact on other species. The most efficient implementation of assimilation is when it is used to minimize a variable that respects the following constraints:

– Its representativity: with greater representativity, the assimilation of a variable is more representative of the size of a model mesh, and therefore it is more accurate.

– Its lifetime: the longer the variability of the variable, the more durably the assimilation will mark the model trajectory, and with fewer errors.

– Its impact: the more the variable impacts the other variables, the more the assimilation gain will propagate toward the other variables that depend on it.

Logically, the earliest studies on chemical assimilation focused on species that had a long lifetime and were directly measurable. As satellite data offered greater spatial coverage, studies were first carried out on species such as CO (**Clerbaux et al. 2001**) on a global scale. Over the years, more and more satellites made it possible to measure more chemical species, such as ozone, NO_2, HCHO and SO_2 (**Martin 2008**). **Adhikary et al. (2008)** assimilated AOD satellite data to optimize aerosol concentration maps in Asia. As the AOD was already used diagnostically, this type of assimilation added uncertainty to the system with respect to account for a concentration **(Saide et al. 2014)**

In parallel to this, studies were carried out on ozone and using surface measurements. **Elbern et al. (1997)** carried out VA using the EURAD model over Europe, **Blond et al. (2003)** used the CHIMERE model and an OI approach over the Parisian region and **Hanea et al. (2004)** used the EUROS model and a Kalman filter. For these studies, the observation operator, which made it possible to "translate" the observations to be assimilated into the model space, becomes a modeling challenge (**Stajner et al. 2004**). Poor representation of the observations in the model space is one of the causes for difficulty in the convergence of the assimilation. An a posteriori application, reconstructing accurate maps, may be an important issue in public health, as in the study by **Zheng et al. (2007)**: data assimilation makes it possible to better reconstruct the transport of a plume from a nuclear accident, and therefore to get a more accurate assessment of the exposure of a population. **Denby et al. (2008)** assimilated PM_{10} data on the surface using the LOTOS-EUROS model. They did this by comparing two different methods: OI and the ensemble Kalman filter (EnKF). They showed that for this pollutant they obtained the best results with OI. **Wu et al. (2008)** also compared assimilation approaches for surface ozone. They also showed that OI gives better results for ozone compared to EnKF and 4D-VAR for the same initial conditions. Then, as the simulation advances, the better the results from EnKF.

8.1.4. *Gain from data assimilation for pollution*

From the time that model and measurements were hybridized, assimilation has always been an asset for creating analysis maps. With regard to forecasting, it is clear that since the system is nonlinear, any improvement in the initial conditions is also a gain. We thus have a system that is globally advantageous, but at the cost of large algorithmic developments: not all direct models for meteorology and chemistry-transport have equivalent adjoint models for all processes.

The question that remains for forecasting through assimilation is: *There is definitely a gain, but for how long? Depending on the domain of study, the resolution and the species being studied, for how many hours will the simulation with assimilation give better results than the simulation without assimilation?*

Blond and Vautard (2004) used OI to assimilate surface ozone data for a forecast. The gain on the error is close to 24 h for a domain over the Paris region. *Tombette et al. (2009)* used OI to assimilate PM_{10} concentrations over France and showed a clear improvement in the simulated concentrations on the first day of a prediction. **Vira and Sofiev (2012)** used 4D-VAR to optimize the initial conditions of their forecasting model and for SO_x concentrations: the study made it possible to refine the emitted mass following the eruption of Etna, as there is always great uncertainty regarding emitted volcanic fluxes when modeling volcanic eruptions. This thus made it possible to improve the simulation for short-term forecasting. More recently, *Menut and Bessagnet (2019)* used academic cases to quantify this gain. Since academic cases are systems that are entirely constrained, this made it possible to have a precise quantification of the gain. For ozone and NO_2, they showed that the maximum gain was of the order of 60 h and 21 h, respectively (Figure 8.3). This gain is after considering that the model has no errors, and that the observations are numerous and representative. This is thus the maximum limit of that which we can never achieve, regardless of models and available measurement assimilation techniques. *Bessagnet et al. (2019)* used the same approach and showed that the gain for PM_{10} could reach 80 h, one part of this gain being that the assimilation corrects emissions for the primary part of this species.

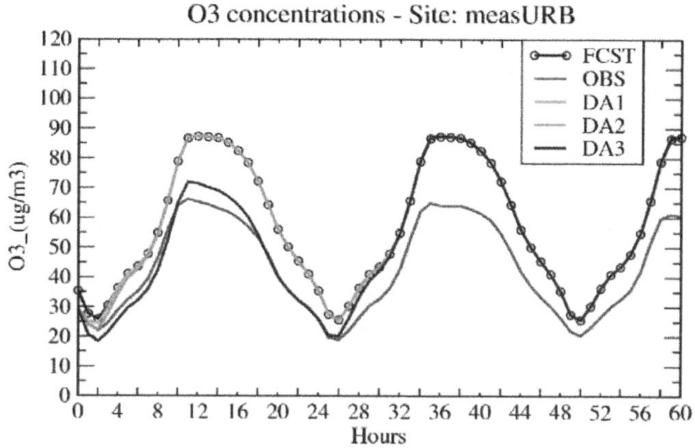

Figure 8.3. *Time series for surface ozone concentrations with different data assimilation scenarios. The "forecast" is the prediction without assimilation. DA1 is the assimilation of only a few surface stations, DA2 is the assimilation of vertical profiles and DA3 is the assimilation of the entire domain, as if it were possible to have information on a species over the entire atmosphere. Figure taken from* Menut and Bessagnet (2019). *For a color version of this figure, see: http://www.iste.co.uk/menut/modeling.zip*

Data assimilation complements modeling and makes it possible to hybridize measurements and the model. Given the nonlinearities in meteorology, this becomes a powerful tool to create a posteriori analyses. Since chemistry-transport is less nonlinear, the gains here are present but smaller. While the gain in analyzing situations a posteriori is clear, there is a smaller gain in forecasting. There is a limit per species, mainly because chemistry-transport is a boundary condition problem (emissions) even more than initial conditions. Depending on assimilation techniques, the effort of developing very complex models (as with adjoints for 4D-VAR) is not clearly visible with respect to the small gain in the results. Further, assimilation is simply a "trajectory correction" for the model and contributes no physical-chemistry information.

8.2. Inverse modeling

Inverse modeling uses the same type of algorithm as used for the analysis, but to a different end – it involves assimilating data over the course of a simulation to optimize a model forcing parameter. As with analysis, the earliest studies were carried out in meteorology and then for chemistry-transport (**Enting 2002**). In chemistry-transport, the studies focused on the inversion of emissions of chemical species (**Reeves et al. 2004**). At first, they were carried out for species with a low lifetime by inversion of the emission of a species with respect to its own concentration by considering transport to be negligible and for the inversion of natural emissions (predictable with meteorology). We see global inversions for methane **(Houweling et al. 1999)**, carbon dioxide **(Bousquet et al. 1999; Kaminski and Heimann 2001; Gurney et al. 2005)**, chlorofluorocarbons (CFCs) **(Mahowald et al. 1997)** and carbon monoxide **(Bergamaschi et al. 2000; Pétron et al. 2004; Müller and Stavrakou 2005)**. On a continental scale, we can see studies on nitrogen oxides **(Konovalov et al. 2005)**. Studies on a local scale are rarer. There are many studies on Atlanta city, the earliest to use inversion of isoprene **(Chang et al. 1996)**, carbon monoxide **(Chang et al. 1997)** and precursors of ozone such as VOCs and NO_x **(Mendoza-Dominguez and Russell 2001)**. **Mulholland and Seinfeld (1995)** studied carbon monoxide in the Los Angeles region. **Quélo et al. (2005)** inverted NO_x emissions in Lille (France). This shows that when the measured species is the same as the inverted species, uncertainties are reduced as there is no chemistry to be considered. Similarly, the lower the resolution of the model, the more constrained the transport, with a smaller source of error in the inversion **(Bousquet et al. 1999; Heimann and Kaminski 1999)**. The mass balance approach is then used **(Enting 2002)**. On a local scale, the emissions are not a continuous process across space or time, making inversion trickier.

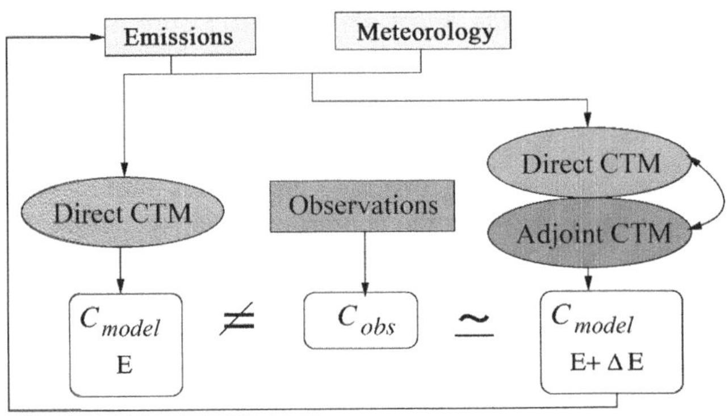

Figure 8.4. *General principle behind inverse modeling. CTM = chemical transport model. For a color version of this figure, see: http://www.iste.co.uk/menut/modeling.zip*

The general principle underlying the calculation of inverse modeling for estimating sources is explained in Figure 8.4. Depending on the emission forcings and meteorology, the simulation makes it possible, first of all, to obtain a model concentration $c_{mod,E}$. These concentrations are different from the observed concentrations c_{obs}. Inverse modeling will successively integrate the direct model and the adjoint model. For each temporal integration, the difference between the model and the observations will be calculated and we will try to minimize the cost function that represents this difference. This minimization will be carried out on all the hours preceding the simulation; by changing emissions over many preceding hours, we optimize the trajectory of the model and obtain a concentration, $c_{mod,E+\Delta E}$, that is closer to the observations.

The academic case makes it possible to bound the solutions. For inverse modeling, an academic case can be created to find out, from the very start, what we wish to find out and whether inversion will help arrive at this. An inversion was carried out in **Pison et al. (2006)** over the Parisian region, with meteorology and emissions considered to be known. This showed that it was possible to once again arrive at a maximum of 80–90% if the source is known. Even in a highly constrained case, this showed that as there were many uncertainties and nonlinearities the exact solution may not necessarily be found. Applying these methods to real cases will probably make it possible to get closer to the solution without determining it exactly.

Using this method does, however, make it possible to greatly improve known emissions or to estimate unknown emissions. This is the case with the massive discharge of pollutants as with the disaster at the Chernobyl nuclear power plant *(Davoine and Bocquet 2007)*, the ETEX release experiment *(Krysta et al. 2008)* or the Fukushima disaster in Japan *(Saunier et al. 2013)*. **Pison et al. (2007)** inverted anthropogenic emissions in the Parisian region and showed that it was necessary to add variability per weekday for traffic and that residential emissions were probably over-estimated in Paris and under-estimated in the suburbs. Inverse modeling can also be used in the long term to reduce uncertainty in trends. *Bergamaschi et al. (2018)* and *McNorton et al. (2018)* showed, through the inter-comparison of methods and models, that it was possible to refine sources and trends in Europe and better attribute the origin of the changes based on activity sector. It is also possible for other parameters to be inverted, as in *Partridge et al. (2012)*, who inverted parameters in a scheme describing cloud-aerosol interactions.

Inverse modeling is a powerful tool for estimating input parameters of a model more accurately. In atmospheric pollution, this makes it possible to better constrain emissions, which are a key factor and among the most uncertain. However, this requires complex tools (adjoint models, minimizer) that are hard to develop and then use. It has also been shown that we do not necessarily arrive at the best solution to the problem. In fact, since minimization considers that all the other parameters and the model need not be optimized, all the errors and uncertainties are carried over onto the inverted parameter. Therefore, while this is satisfactory for a given model and configuration, an inverted variable will not necessarily be appropriate for another model, another resolution or a different period being studied.

8.3. Forecasting

Open-access publication: *Marécal et al. (2015)*.

Forecasting remains one of the major goals in the development of numerical tools dedicated to estimating concentration ranges of pollutants. If we wish to take steps to rapidly reduce a pollution peak that is predicted a few days in advance, we must develop models that are both accurate and stable in behavior (*Menut and Bessagnet 2010*; **Zhang et al. 2012a**). Further, while forecasting is often considered to be the end-result of developing and using a model, it is also the best means of objectively evaluating a tool's day-to-day abilities. Forecasting may take place over different timescales. For regional modeling, it is often a few days in advance, with the major constraint being the pertinence of the meteorological predictions that will guide the chemistry-transport predictions. Forecasting uses a wide variety of models, not only deterministic models. Consequently, systems for detecting pollution peaks are based on statistical approaches like neural networks (**Robeson and Steyn 1990**). For example, **Lightstone et al. (2017)** compared the CMAQ model with a neural network for New York City and found that the results from the neural network and the CTM were of the same quality. This does not, however, hold true for "atypical events". For a neural network to detect an unusual event, such an event should already have occurred and been learned by the network.

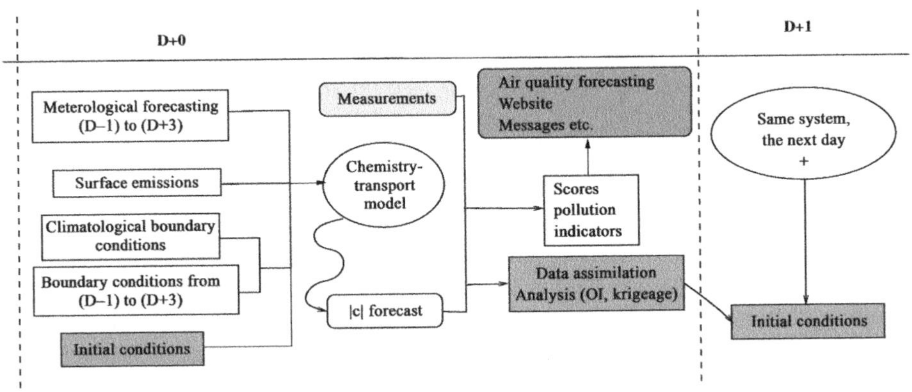

Figure 8.5. *General principle behind an air quality forecasting chain. For a color version of this figure, see: http://www.iste.co.uk/menut/modeling.zip*

Figure 8.5 depicts a diagram showing the chain of modeling operations for a daily forecast. The principle here harks back to the classical analysis simulation, but with the addition of "chaining". There must be a series of simulations day after day, and the simulation for day (D+1) must be initialized using the simulation carried out on day (D). For instance, the forcings required to predict the chemistry-transport over the period (D-1) to (D+3) are as follows: (i) meteorology, emissions and regional boundary conditions for this period; (ii) the initial conditions for the start of the forecast.

There are many forecasting systems, whether experimental (for measurement campaigns, or to test models) or operational. Operational systems are implemented either by meteorological centers or by dedicated bodies, such as associations that monitor air quality. The earliest operational forecasting systems emerged around the 2000s **(Vautard et al. 2001; Delle Monache and Stull 2003; McHenry et al. 2004)**. Following this, other systems were implemented as described in journal articles by *Menut and Bessagnet (2010)*, **Zhang et al. (2012a)**, and **Zhang et al. (2012b)**. In France, there was additional structuring with the PREVAIR forecasting system put in place by INERIS using the CHIMERE model (CNRS/INERIS) and the MOCAGE model (Météo-France) **(Honoré et al. 2008)**. For Europe, the CAMS laboratory consortium *(Marécal et al. 2015)*, part of the COPERNICUS program, carried out a daily forecast through the European Centre for Medium-Range Weather Forecasts (ECMWF). For meteorology, there was the IFS forecasting model **(Morcrette et al. 2009)**.

Among all the uncertainties related to forecasting, the question that arises is *how can forecasting be improved?* We encounter the same constraints as in analysis: the quality of parametrizations, the number of processes represented in the model, the spatial and temporal resolution of the model, the impact of the initial conditions and the forcings, etc. In addition, there are specific constraints: the model must be quick and produce a forecast in a few hours; it must also be numerically stable as the forecasting must function every day. Given that the impact of the initial conditions is important, different techniques are tested, as in **Blond and Vautard (2004)**, who assimilated surface data to provide ranges for ozone that were closest to reality at the start of the simulation. Ensemble techniques were also used – several forecasts yielded better scores than independent simulations *(Bei et al. 2010; Marécal et al. 2015; Brasseur et al. 2019)*.

Forecasting may also be a tool for validating a model. By comparing forecasts day after day, it is possible to evaluate the scattering of the obtained results based on changes in initialization. For example, *Menut et al. (2015b)* demonstrated the impact of a wind error on other processes that depended on this: a small change in the wind speed would change mineral dust emissions, and the direction of the wind would then change the transport of this plume. Deposition would then be modified (depending on the available mass and on the wind). Finally, surface concentrations of pollutants would differ between the forecasting periods. Thus, comparing these different periods is an indicator of the variability of the model. Figure 8.6 depicts maps showing the differences in PM_{10} concentrations between multiple simulations: (D + 1) - (D + 0) and (D + 3) - (D + 0). It can be clearly seen that the difference increases with the forecasting timeframe, providing a quantification of the uncertainty in the forecast.

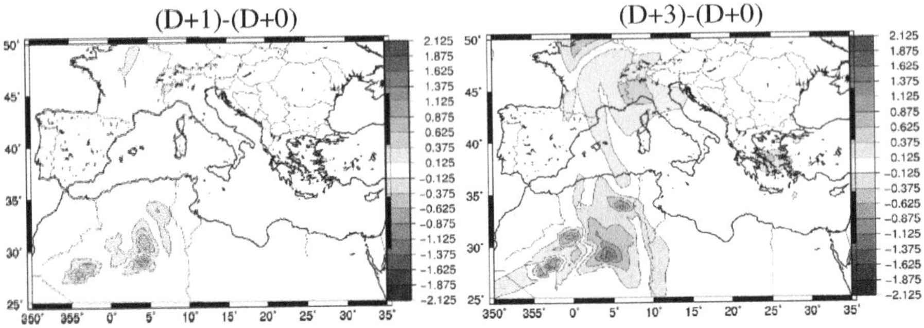

Figure 8.6. *Variability in the forecasts of PM_{10} ($\mu g\ m^{-3}$) on the surface, depending on the timeframe. Figure adopted from Menut et al. (2015b). For a color version of this figure, see: http://www.iste.co.uk/menut/modeling.zip*

Forecasting remains one of the goals of modeling atmospheric composition. The results of the forecast are highly dependent on nonlinearities in meteorology and, therefore, on their strong dependence on the initial condition. They also depend on the quality of the model and the parametrizations, which must be constantly improved. The ways to limit these uncertainties are carrying out ensemble forecasts, as well as initializing the model using ranges that have benefited from data assimilation. Certain forcings will still remain a very rough estimate, such as the forecasting of road traffic in a large city, or the duration of a wildfire observed by satellite.

9

The Impacts of Atmospheric Pollution

This chapter looks at how atmospheric pollution impacts the two broad themes justifying its study: climate (including meteorology and vegetation) and health. We will study the following questions:

1) Can regional models quantify the impact of pollution on climate and of climate on pollution, and impact of climate and pollution on health?

2) Can long-term trends be modeled regionally?

3) Which are the impacts being modeled at present? What other impacts are yet to be quantified and what are the developments required in the models?

9.1. The impact of pollution on climate

Recent climate change is the result of the excessive atmospheric pollution from the time of the industrial period. Therefore, the impact on the climate is quite clear, since pollution is the cause for this change. Pollution and climate are, in fact, the same phenomenon on different spatial and temporal scales. Atmospheric pollution is a phenomenon with high temporal frequency (ranging from a minute to a year), with a spatial coverage that is often local (a city) or regional. On the other hand, climate is a phenomenon with low temporal frequency (from a year to a millennium) with global coverage. However, in both cases these systems are affected by important anthropogenic actions. It is difficult to quantify this impact and models are used as support to analyze long periods in the past and to make future projections. From the point of view of chemical species, there is no direct correspondence between the two, as there may be different species, as shown in Table 9.1. However, the common point is excessive anthropogenic emissions (**Jacobson and Streets 2009**). With respect to climate, attention is largely focused on greenhouse gases. In the long term, it is the radiative impact of these gases that will modify climate, especially CO_2 and CH_4.

Laurent MENUT. © ISTE Ltd 2024.

With respect to air quality we see pollutants whose tracking is governed by regulations, with a shorter lifetime, but whose impact on health and vegetation is recognized.

Principal greenhouse gases	Principal AQ pollutants
Water vapor (H_2O)	Ozone (O_3)
Carbon dioxide (CO_2)	Nitrogen oxides d'azote (NO_x)
Methane (CH_4)	Sulfur dioxide (SO_2)
Ozone (O_3)	Volatile organic compounds (VOCs)
Nitrous oxide (N_2O)	Polycyclic aromatic hydrocarbons (PAH)
(Hydro-) Chlorofluorocarbons ((H-)CFC)	$PM_{2.5}$ and PM_{10} particles
Sulfur hexafluoride (SF_6)	Ammoniac (NH_3), Benzene (C_6H_6) Carbon monoxide (CO)

Table 9.1. *Comparison between chemical species involved in the "climate" and "air quality" systems*

Although climate and pollution are not on the same spatial-temporal scales, certain comparisons have been carried out on a regional scale, over a few decades, in the context of *regional climate* studies. Pollution can notably rapidly impact climate change through the over-concentration of certain gases and aerosols *(Zhao et al. 2019)*. Gases, especially greenhouse gases (CO_2, N_2O, CH_4), will absorb infrared radiation, causing the atmosphere to heat up. Ozone at altitude will modify the oxidizing capacity of the atmosphere. Aerosols have a radiative forcing that depends on their size distribution and, therefore, on their diffusive or absorbent effect **(Haywood and Boucher 2000)**. For example, mineral dust will have a negative radiative forcing for short-wave (solar) radiation, because they will diffuse the radiation and this happens for the fine part of the aerosol (0.1–1 μm). However, mineral dust will have a positive radiative impact for long-wave (infrared) radiation by absorbing radiation, and this happens with the coarser part of the aerosol (5–100 μm) **(Mahowald et al. 2011)**. Since atmospheric pollution is at the origin of climate change, the solutions are theoretically obvious and involve a reduction of anthropogenic emissions *(Turnock et al. 2018)*. However, putting this into practice is difficult: for each sector of activity that must reduce emissions, we must find a nonpolluting equivalent. And in a few cases, the modeling of reduced emissions may show a smaller impact than hoped for *Leibensperger et al. (2012)*.

Regional modeling cannot accurately quantify the impact of pollution on the climate as this involves very long timescales. This implies that using global models or ensemble models for atmospheric circulation may be considered appropriately. At the same time, "regional climate" has developed greatly as a topic, simulating several decades but still sensitive to the accuracy of the boundary conditions when it comes to studying trends.

9.2. The impact of climate on pollution

While the impact of pollution on climate is evident, climate change also conversely impacts pollution. This must be taken into account if we wish to accurately model regional pollution, especially when carrying out calculations for scenarios or for trends over several years **(Jacob and Winner 2009)**. "Modeling regional climate" began to be developed a few decades ago in order to model these interactions **(Beniston et al. 2007)**; **(Giorgi 2019)**. These CTMs are conceptually different from the CTMs used for case studies and short-term forecasting: they do not try to obtain the best quantification of a concentration at a given site, but must carry out simulations very rapidly and accurately describe variabilities and trends. There are many uncertainties around the modeled projections and projects such as PRUDENCE aim to quantify these **(Déqué et al. 2007)**. They showed that for regional models, forcing at the boundaries is primordial, especially for temperature. Ensemble modeling remains the best-adapted tool for gauging the pertinence of these simulations for the long term *(Nikulin et al. 2011)*.

The impact of climate change can be seen in two aspects: meteorology and vegetation. Since atmospheric chemistry depends directly on meteorology, it is expected that climate change will have an impact **(Jacob and Winner 2009)**. A rise in temperature will lead to an increase in photochemical activity and, consequently, to a rise in ozone concentrations **(Meleux et al. 2007)**. On the contrary, an expected increase in cloud cover may limit this photochemistry. **Fang et al. (2011)** showed that the fall in precipitation in certain future climate scenarios will lead to a decrease in wet deposition and, consequently, to an increase in surface aerosol concentrations. **Katragkou et al. (2011)** confirmed the increase in ozone in Europe, being also due to a modification in the anticyclone in Western Europe and therefore due to many episodes of stagnation.

As the impacts are highly dependent on the accuracy of the future climate scenarios, we must question the quality of these global climate models. The difficulty, of course, is the fact that comparison data are not available. We must therefore compare simulations, created with these climatic tools, over past periods and with measurements. **Menut et al. (2013b)** compared regional simulations with climatic simulations and showed that compensations for model errors made interpretation difficult. In summer, the climatic model will have a positive bias for solar radiation, but a negative bias for the surface wind speed. This induces too many stagnation episodes and a negative temperature bias. In winter, while the dynamic shows no systematic bias, the precipitations are overestimated, which then leads to an underestimation of the future aerosols predicted. Figure 9.1 presents two rain maps, with biases, calculated based on the forcing climatic model. It can be seen that the climate model may induce a much greater model than the regional model that is usually used for pollution studies.

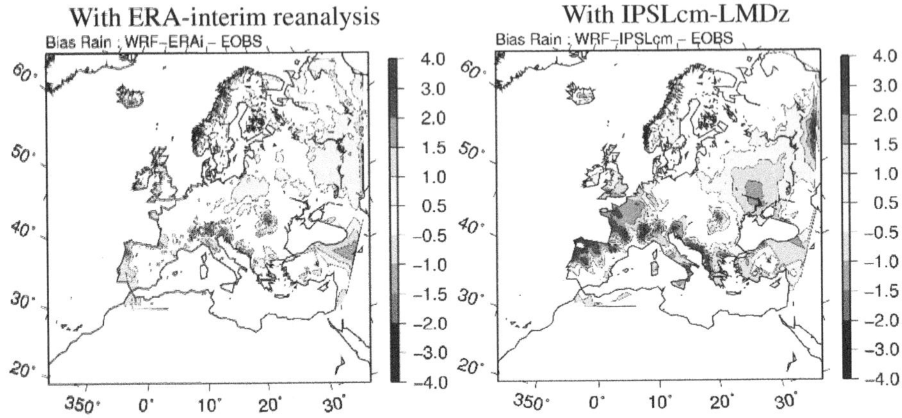

Figure 9.1. *Bias in precipitations from 1989 to 2005 depending on the meteorological forcing. As per data from* **Menut et al. (2013b)**. *For a color version of this figure, see: http://www.iste.co.uk/menut/ modeling.zip*

Climate change also has an impact on vegetation, which in turn will impact atmospheric pollution. *Wang et al. (2020)* thus showed that reforestation may locally induce a direr and warmer climate, inducing higher ozone concentrations. The feedback mechanisms are complex, but as with short-term regional pollution studies, it is seen that we must take into account a meteorology-chemistry-vegetation coupling to effectively consider all of the processes.

The difficulty with these studies is the accumulation of errors over time. The regional model is relatively precise and uses parametrizations that are quite fine and thus variable enough. Its schemes are more sensitive and thus more uncertain than robust schemes from climate models. While their use is recommended for short simulations (a few days to a few months), using them over decades may cause large biases to appear, with errors accumulating over time.

9.3. The impact of pollution on health

One of the goals of modeling pollution is to look at the impact pollution has on health. However, regional modeling only brings in partial information as the estimation of this impact is a multidisciplinary field involving geophysicists, sociologists and epidemiologists. For more on these subjects, we recommend publications that are available through the website of the Observatoire Régional de la Santé (ORS) (Regional Health Observatory: https://www.ors-idf.org/). The article by *Malmqvist et al. (2018)* describes all recent programs that have studied these impacts.

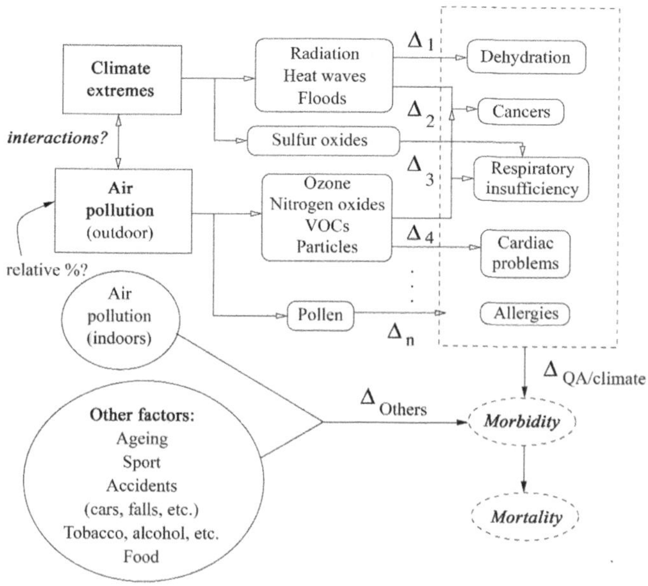

Figure 9.2. *Diagram summarizing the various links between chemical concentrations of pollutants and illnesses*

The regional modeling of pollution will thus provide an answer to the first link in the chain of impacts. This chain can be summarized as depicted in Figure 9.2:

– a type of pollution or modification in meteorology: climate, indoor pollution, outdoor pollution, etc.;

– a type of impact on the meteorology: radiation, overconcentration of a chemical species, pollen, etc.;

– a type of health problem: cardiac problems, cancer, allergy, etc.;

– an incidence on morbidity, and then on mortality. Certain studies have also quantified the economic cost of this impact *(Im et al. 2018a)*.

The pollution model only contributed data on atmospheric concentrations. There are mostly just as many uncertainties for the other kinds of quantification related to the impact. With regard to pollution, we also note that this impact can only be quantified for the "outdoor" contribution. There are also studies on the impact of indoor pollution, but an analysis with both simultaneously remains marginal. For example, **Brinkman et al. (2009)** measured chemical concentrations and tried to posit a speciation to differentiate between indoors and outdoors. In terms of health impact, and to have a complete picture, we must also take into account food, water quality, etc. At present, by considering only outdoor pollution, we have a very fragmented view of the impact that pollution has.

Although all types of pollution may impact health, the anthropogenic part is the most studied as it is the easiest to reduce **(Anderson 2009)**. Among all pollutants, particles have been recognized, for a few decades now, as having a negative impact on health *(Kim et al. 2015)*. Outdoor and urban PM_{10} particles have an estimated effect of 0.8 million premature deaths per year, and indoor combustion of fuel is associated with 2 million premature deaths worldwide. While the health impact remains a priority, many studies have also tried to quantify the cost and benefits of high aerosol concentrations with respect to health, vegetation and the climate **(Löndahl et al. 2010)**. Table 9.2 summarizes a few of the positive and negative effects of aerosol reduction in the atmosphere. It is clear that the majority of effects of climate change are negative, including the effects on the abundance of aerosols and the feedback effect this increase has on health and the climate.

Climate change	Effect on health	Effect of aerosols
Rising temperature	↘ More allergens ↗ More vegetation ↘ Degradation of food ↘ Increase in water-borne diseases	More aerosols ↗ Cooling effect
Flooding	↘ Habitat destruction (deaths) ↘ Transport of toxic products, illnesses	↘ + Intense rains
Drought	↗ Lower transport of diseases ↘ Decrease in water availability ↘ Reduction in vegetation	↘ Less precipitation

Table 9.2. *Impact of aerosols depending on climate and health. The symbols ↗ and ↘ denote the positive and negative effects, respectively (adopted from* **Löndahl et al. (2010)***)*

There are specific cases to simplify the quantification of impacts and successfully correlate the causes and effects. We may have an intense pollution event that makes it possible to eliminate all other possible and habitual cause. For example, a volcanic eruption, a heat-wave or recurrent "abnormal" pollution (e.g. Beijing and China). It is also possible to work on events that cause an isolated and characteristic health problem: for example, pollen and the associated respiratory problems or a modification in the UV index and its impact *(Jégou et al. 2011)*.

Regional modeling can contribute by participating in the construction of a **risk model**. This promotes collaborations between geophysicists and epidemiologists. This risk model also includes an **exposure model**. This is the area where research is developing in the geophysics community. The exposure model will be associated with a chemistry-transport model and a population model. In order to associate both of these, the chemistry-transport model must have the best possible spatial resolution **(Gaines Wilson and Zawar-Reza 2006)**; **(Thompson and Selin 2012)**. One of the earliest research projects on this topic was the FUMAPEX project in Helsinki

(Baklanov et al. 2007). We can also note, among the earliest studies, the work carried out in Athens, Greece (**Kassomenos et al. 2008**), in Valencia, Spain (**Viana et al. 2008**) and Portugal (**Borrego et al. 2009**). To avoid a very costly drop in resolution, techniques have also been developed for subgrid variability in chemical concentrations based on activity sectors associated with anthropogenic emissions and this makes it possible to supply exposure models (**Valari et al. 2011**).

Impact studies carried out for current periods are also accompanied by studies on future periods. Notably, climate change is also studied to find out whether this will modify the health impact, as well as the associated cost, depending on scenarios (*Colette et al. 2012*; *Brandt et al. 2013*; *Tarín-Carrasco et al. 2019*). In Europe, it was concluded that in general, the number of diseases would increase by a factor of two by the end of the century, increasing health costs similarly.

Studies on the impact of pollution on health are growing now. These studies are one of the clear end-goals of pollution studies. However, in order to develop they had to wait for models of atmospheric pollution to reach a minimum level of maturity so that the results would make sense. Collaborations have recently been established between geophysicists and epidemiologists to advance studies in this field. However, we still have significant uncertainties with respect to geophysics, the exposure of populations and the risk that this exposure poses to the populations. Furthermore, since it is not only outdoor pollution that can affect human health, we must also take into account indoor pollution, food, toxicity of medication, etc. We are therefore still quite a long way from achieving accurate quantification of this impact and the current studies that provide a number of deaths, or put a figure to economic costs, are more than uncertain.

References

Abdul-Razzak, H. and Ghan, S.J. (2002). A parameterization of aerosol activation 3. Sectional representation. *Journal of Geophysical Research: Atmospheres*, 107(D3), AAC 1–1–AAC 1–6.

Adhikary, B., Kulkarni, S., Dallura, A., Tang, Y., Chai, T., Leung, L., Qian, Y., Chung, C., Ramanathan, V., Carmichael, G. (2008). A regional scale chemical transport modeling of Asian aerosols with data assimilation of AOD observations using optimal interpolation technique. *Atmospheric Environment*, 42(37), 8600–8615.

Al-Wali, K. and Samson, P. (1996). Preliminary sensitivity analysis of urban airshed model simulations to temporal and spatial availability of boundary layer wind measurements. *Atmospheric Environment*, 30, 2027–2042.

Albrecht, R.I., Goodman, S.J., Buechler, D.E., Blakeslee, R.J., Christian, H.J. (2016). Where are the lightning hotspots on earth? *Bulletin of the American Meteorological Society*, 97(11), 2051–2068.

Alfaro, S.C. and Gomes, L. (2001). Modeling mineral aerosol production by wind erosion: Emission intensities and aerosol size distribution in source areas. *Journal of Geophysical Research*, 106, 18075–18084.

Alfaro, S.C., Gaudichet, A., Gomes, L., Maillé, M. (1998). Mineral aerosol production by wind erosion: Aerosol particle sizes and binding energies. *Geophysical Research Letters*, 25, 991–994.

Allen, D.J. and Pickering, K.E. (2002). Evaluation of lightning flash rate parameterizations for use in a global chemical transport model. *Journal of Geophysical Research: Atmospheres*, 107(D23), ACH 15–1–ACH 15–21.

Allen, D.J., Pickering, K., Pinder, R.W., Pierce, T. (2009). Impact of lightning-NO emissions on eastern US photochemistry during the summer of 2004 as determined using the CMAQ model. In *8th Annual CMAS Conference*, October 19–21. Chapel Hill.

Allen, D.J., Pickering, K.E., Pinder, R.W., Henderson, B.H., Appel, K.W., Prados, A. (2012). Impact of lightning-NO on eastern United States photochemistry during the summer of 2006 as determined using the CMAQ model. *Atmospheric Chemistry and Physics*, 12(4), 1737–1758.

Amundson, N.R., Caboussat, A., He, J.W., Martynenko, A.V., Savarin, V.B., Seinfeld, J.H., Yoo, K.Y. (2006). A new inorganic atmospheric aerosol phase equilibrium model (UHAERO). *Atmospheric Chemistry and Physics*, 6(4), 975–992.

Amundson, N.R., Caboussat, A., He, J.W., Martynenko, A.V., Landry, C., Tong, C., Seinfeld, J.H. (2007). A new atmospheric aerosol phase equilibrium model (UHAERO): Organic systems. *Atmospheric Chemistry and Physics*, 7(17), 4675–4698.

Anderson, H. (2009). Air pollution and mortality: A history. *Atmospheric Environment*, 43(1), 142–152.

Andreae, M. and Rosenfeld, D. (2008). Aerosol-cloud-precipitation interactions. Part 1. The nature and sources of cloud-active aerosols. *Earth-Science Reviews*, 89, 13–41.

Appel, K.W., Foley, K.M., Bash, J.O., Pinder, R.W., Dennis, R.L., Allen, D.J., Pickering, K. (2011). A multi-resolution assessment of the community multiscale air quality (CMAQ) model v4.7 wet deposition estimates for 2002–2006. *Geoscientific Model Development*, 4(2), 357–371.

Archer-Nicholls, S., Lowe, D., Schultz, D.M., McFiggans, G. (2016). Aerosol–radiation–cloud interactions in a regional coupled model: The effects of convective parameterisation and resolution. *Atmospheric Chemistry and Physics*, 16(9), 5573–5594.

Atger, F. (1999). The skill of ensemble prediction systems. *Monthly Weather Review*, 127, 1941–1953.

Auger, L. and Legras, B. (2007). Chemical segregation by heterogeneous emissions. *Atmospheric Environment*, 41(11), 2303–2318.

Aumont, B., Szopa, S., Madronich, S. (2005). Modelling the evolution of organic carbon during its gas-phase tropospheric oxidation: Development of an explicit model based on a self generating approach. *Atmospheric Chemistry and Physics*, 5(9), 2497–2517.

Baik, J. and Chun, H. (1997). A dynamical model for urban heat islands. *Boundary-Layer Meteorology*, 83, 463–477.

Baklanov, A. (2006). Overview of the European project FUMAPEX. *Atmospheric Chemistry and Physics*, 6(7), 2005–2015.

Baklanov, A., Hänninen, O., Slørdal, L.H., Kukkonen, J., Bjergene, N., Fay, B., Finardi, S., Hoe, S.C., Jantunen, M., Karppinen, A. et al. (2007). Integrated systems for forecasting urban meteorology, air pollution and population exposure. *Atmospheric Chemistry and Physics*, 7(3), 855–874.

Baklanov, A., Mestayer, P.G., Clappier, A., Zilitinkevich, S., Joffre, S., Mahura, A., Nielsen, N.W. (2008). Towards improving the simulation of meteorological fields in urban areas through updated/advanced surface fluxes description. *Atmospheric Chemistry and Physics*, 8(3), 523–543 [Online]. Available at: acp.copernicus.org/articles/8/523/2008/.

Baklanov, A., Schlünzen, K., Suppan, P., Baldasano, J., Brunner, D., Aksoyoglu, S., Carmichael, G., Douros, J., Flemming, J., Forkel, R. et al. (2014). Online coupled regional meteorology chemistry models in Europe: Current status and prospects. *Atmospheric Chemistry and Physics*, 14(1), 317–398.

Baklanov, A., Smith Korsholm, U., Nuterman, R., Mahura, A., Nielsen, K.P., Sass, B.H., Rasmussen, A., Zakey, A., Kaas, E., Kurganskiy, A. et al. (2017). Enviro-HIRLAM online integrated meteorology–chemistry modelling system: Strategy, methodology, developments and applications (v7.2). *Geoscientific Model Development*, 10(8), 2971–2999.

Baldauf, M., Seifert, A., Förstner, J., Majewski, D., Raschendorfer, M., Reinhardt, T. (2011). Operational convective-scale numerical weather prediction with the COSMO model: Description and sensitivities. *Monthly Weather Review*, 139, 3887–3905.

Baldridge, K. and Cox, W. (1986). Evaluating air quality model performance. *Environmental Software*, 1(3), 182–187.

Balkanski, Y., Schulz, M., Claquin, T., Guibert, S. (2007). Reevaluation of mineral aerosol radiative forcings suggests a better agreement with satellite and AERONET data. *Atmospheric Chemistry and Physics*, 7(1), 81–95.

Banerjee, A., Archibald, A.T., Maycock, A.C., Telford, P., Abraham, N.L., Yang, X., Braesicke, P., Pyle, J.A. (2014). Lightning no_x, a key chemistry-climate interaction: Impacts of future climate change and consequences for tropospheric oxidising capacity. *Atmospheric Chemistry and Physics*, 14(18), 9871–9881.

Bangert, M., Kottmeier, C., Vogel, B., Vogel, H. (2011). Regional scale effects of the aerosol cloud interaction simulated with an online coupled comprehensive chemistry model. *Atmospheric Chemistry and Physics*, 11(9), 4411–4423.

Barlow, J.F. (2014). Progress in observing and modelling the urban boundary layer. *Urban Climate*, 10, 216–240.

Barlow, J.F., Rooney, G., von Hunerbein, S., Bradley, S. (2008). Relating urban surface-layer structure to upwind terrain for the Salford experiment (Salfex). *Boundary-Layer Meteorology*, 127, 173–191.

Baró, R., Palacios-Peña, L., Baklanov, A., Balzarini, A., Brunner, D., Forkel, R., Hirtl, M., Honzak, L., Pérez, J.L., Pirovano, G. et al. (2017). Regional effects of atmospheric aerosols on temperature: An evaluation of an ensemble of online coupled models. *Atmospheric Chemistry and Physics*, 17(15), 9677–9696.

Barret, B., Williams, J.E., Bouarar, I., Yang, X., Josse, B., Law, K., Pham, M., Le Flochmoën, E., Liousse, C., Peuch, V.H. et al. (2010). Impact of West African monsoon convective transport and lightning no_x production upon the upper tropospheric composition: A multi-model study. *Atmospheric Chemistry and Physics*, 10(12), 5719–5738.

Barthe, C. and Barth, M.C. (2008). Evaluation of a new lightning-produced no_x parameterization for cloud resolving models and its associated uncertainties. *Atmospheric Chemistry and Physics*, 8(16), 4691–4710.

Basart, S., Pérez, C., Cuevas, E., Baldasano, J.M., Gobbi, G.P. (2009). Aerosol characterization in Northern Africa, Northeastern Atlantic, Mediterranean Basin and Middle East from direct-sun AERONET observations. *Atmospheric Chemistry and Physics*, 9(21), 8265–8282.

Baumgarten, G. (2010). Doppler Rayleigh/Mie/Raman lidar for wind and temperature measurements in the middle atmosphere up to 80 km. *Atmospheric Measurement Techniques*, 3(6), 1509–1518.

Beardsley, R.L. and Jang, M. (2016). Simulating the SOA formation of isoprene from partitioning and aerosol phase reactions in the presence of inorganics. *Atmospheric Chemistry and Physics*, 16(9), 5993–6009.

Beegum, S.N., Gherboudj, I., Chaouch, N., Couvidat, F., Menut, L., Ghedira, H. (2016). Simulating aerosols over Arabian Peninsula with CHIMERE: Sensitivity to soil, surface parameters and anthropogenic emission inventories. *Atmospheric Environment*, 128, 185–197.

Beekmann, M. and Vautard, R. (2010). A modelling study of photochemical regimes over Europe: Robustness and variability. *Atmospheric Chemistry and Physics*, 10, 10067–10084.

Beekmann, M., Prévôt, A.S.H., Drewnick, F., Sciare, J., Pandis, S.N., Denier van der Gon, H.A.C., Crippa, M., Freutel, F., Poulain, L., Ghersi, V. et al. (2015). In situ, satellite measurement and model evidence on the dominant regional contribution to fine particulate matter levels in the Paris megacity. *Atmospheric Chemistry and Physics*, 15(16), 9577–9591.

Bei, N., Lei, W., Zavala, M., Molina, L.T. (2010). Ozone predictabilities due to meteorological uncertainties in the Mexico city basin using ensemble forecasts. *Atmospheric Chemistry and Physics*, 10(13), 6295–6309.

Beirle, S., Huntrieser, H., Wagner, T. (2010). Direct satellite observation of lightning-produced no_x. *Atmospheric Chemistry and Physics*, 10(22), 10965–10986.

Benedetti, A., Reid, J.S., Knippertz, P., Marsham, J.H., Di Giuseppe, F., Rémy, S., Basart, S., Boucher, O., Brooks, I.M., Menut, L. et al. (2018). Status and future of numerical atmospheric aerosol prediction with a focus on data requirements. *Atmospheric Chemistry and Physics*, 18(14), 10615–10643.

Beniston, M., Stephenson, D.B., Christensen, O.B., Ferro, C.A.T., Frei, C., Goyette, S., Halsnaes, K., Holt, T., Jylha, K., Koffi, B. et al. (2007). Future extreme events in European climate: An exploration of regional climate model projections. *Climatic Change*, 81(1), 71–95.

van den Berg, A., Dentener, F., Lelieveld, J. (2000). Modeling the chemistry of the marine boundary layer: Sulphate formation and the role of sea-salt aerosol particles. *Journal of Geophysical Research: Atmospheres*, 105(D9), 11671–11698.

Berg, L.K., Shrivastava, M., Easter, R.C., Fast, J.D., Chapman, E.G., Liu, Y., Ferrare, R.A. (2015). A new WRF-Chem treatment for studying regional-scale impacts of cloud processes on aerosol and trace gases in parameterized cumuli. *Geoscientific Model Development*, 8(2), 409–429.

Bergamaschi, P., Hein, R., Heinmann, M., Crutzen, P. (2000). Inverse modeling of the global CO cycle: 1. Inversion of CO mixing ratio. *Journal of Geophysical Research*, 105(D2), 1909–1927.

Bergamaschi, P., Karstens, U., Manning, A.J., Saunois, M., Tsuruta, A., Berchet, A., Vermeulen, A.T., Arnold, T., Janssens-Maenhout, G., Hammer, S. et al. (2018). Inverse modelling of European ch_4 emissions during 2006–2012 using different inverse models and reassessed atmospheric observations. *Atmospheric Chemistry and Physics*, 18(2), 901–920.

Bergamo, A., Tafuro, A.M., Kinne, S., De Tomasi, F., Perrone, M.R. (2008). Monthly-averaged anthropogenic aerosol direct radiative forcing over the Mediterranean based on AERONET aerosol properties. *Atmospheric Chemistry and Physics*, 8(23), 6995–7014.

Bergström, R., Denier van der Gon, H.A.C., Prévôt, A.S.H., Yttri, K.E., Simpson, D. (2012). Modelling of organic aerosols over Europe (2002–2007) using a volatility basis set (VBS) framework: Application of different assumptions regarding the formation of secondary organic aerosol. *Atmospheric Chemistry and Physics*, 12(18), 8499–8527.

Bessagnet, B., Hodzic, A., Vautard, R., Beekmann, M., Cheinet, S., Honoré, C., Liousse, C., Rouil, L. (2004). Aerosol modeling with CHIMERE: Preliminary evaluation at the continental scale. *Atmospheric Environment*, 38, 2803–2817.

Bessagnet, B., Menut, L., Curci, G., Hodzic, A., Guillaume, B., Liousse, C., Moukhtar, S., Pun, B., Seigneur, C., Schulz, M. (2009). Regional modeling of carbonaceous aerosols over Europe – Focus on secondary organic aerosols. *Journal of Atmospheric Chemistry*, 61, 175–202.

Bessagnet, B., Pirovano, G., Mircea, M., Cuvelier, C., Aulinger, A., Calori, G., Ciarelli, G., Manders, A., Stern, R., Tsyro, S. et al. (2016). Presentation of the EURODELTA III intercomparison exercise – Evaluation of the chemistry transport models' performance on criteria pollutants and joint analysis with meteorology. *Atmospheric Chemistry and Physics*, 16(19), 12667–12701.

Bessagnet, B., Menut, L., Couvidat, F., Meleux, F., Siour, G., Mailler, S. (2019). What can we expect from data assimilation for air quality forecast? Part II: Analysis with a semi-real case. *Journal of Atmospheric and Oceanic Technology*, 36, 1433–1448.

Bessagnet, B., Menut, L., Lapere, R., Couvidat, F., Jaffrezo, J.-L., Mailler, S., Favez, O., Pennel, R., Siour, G. (2020). High resolution chemistry transport modeling with the online CHIMERE-WRF model over the French Alps – Analysis of a feedback of surface particulate matter concentrations on mountainous meteorology. *Atmosphere*, 11(6), 565.

Bessagnet, B., Beauchamp, M., Menut, L., Fablet, R., Pisoni, E., Thunis, P. (2021). Deep learning techniques applied to super-resolution chemistry transport modeling for operational uses. *Environmental Research Communications*, 3(8), 085001.

Bicheron, P., Amberg, V., Bourg, L., Petit, D., Huc, M., Miras, B., Brockmann, C., Hagolle, O., Delwart, S., Ranera, F. et al. (2011). Geolocation assessment of MERIS GlobCover orthorectified products. *IEEE Transactions on Geoscience and Remote Sensing*, 49(8), 2972–2982.

Blond, N. and Vautard, R. (2004). Three-dimensional ozone analyses and their use for short-term ozone forecasts. *Journal of Geophysical Research-Atmospheres*, 109(D17). doi: 10.1029/2004JD004515.

Blond, N., Bel, L., Vautard, R. (2003). Three-dimensional ozone data analysis with an air quality model over the Paris area. *Journal of Geophysical Research-Atmospheres*, 108(D23). doi: 10.1029/2003JD003679.

Bocquet, M., Elbern, H., Eskes, H., Hirtl, M., abkar, R., Carmichael, G.R., Flemming, J., Inness, A., Pagowski, M., Pérez Camaño, J.L. et al. (2015). Data assimilation in atmospheric chemistry models: Current status and future prospects for coupled chemistry meteorology models. *Atmospheric Chemistry and Physics*, 15(10), 5325–5358.

Boersma, K.F., Eskes, H.J., Meijer, E.W., Kelder, H.M. (2005). Estimates of lightning NO_x production from GOME satellite observations. *Atmospheric Chemistry and Physics*, 5(9), 2311–2331.

Boichu, M., Menut, L., Khvorostyanov, D., Clarisse, L., Clerbaux, C., Turquety, S., Coheur, P.-F. (2013). Inverting for volcanic SO2 flux at high temporal resolution using spaceborne plume imagery and chemistry-transport modelling: The 2010 Eyjafjallajökull eruption case-study. *Atmospheric Chemistry and Physics*, 13, 8569–8584.

Borrego, C., Sá, E., Monteiro, A., Ferreira, J., Miranda, A. (2009). Forecasting human exposure to atmospheric pollutants in Portugal – A modelling approach. *Atmospheric Environment*, 43(36), 5796–5806.

Bousquet, P., Ciais, P., Peylin, P., Monfray, P. (1999). Inverse modeling and annual atmospheric CO_2 sources and sinks 1: Method and control inversion. *Journal of Geophysical Research*, 104(D21), 26161–26178.

Bozzo, A., Benedetti, A., Flemming, J., Kipling, Z., Rémy, S. (2020). An aerosol climatology for global models based on the tropospheric aerosol scheme in the integrated forecasting system of ECMWF. *Geoscientific Model Development*, 13(3), 1007–1034.

Brandt, J., Silver, J.D., Christensen, J.H., Andersen, M.S., Bønløkke, J.H., Sigsgaard, T., Geels, C., Gross, A., Hansen, A.B., Hansen, K.M. et al. (2013). Contribution from the ten major emission sectors in Europe and Denmark to the health-cost externalities of air pollution using the EVA model system – An integrated modelling approach. *Atmospheric Chemistry and Physics*, 13(15), 7725–7746.

Brasseur, G. and Jacob, D. (2017). *Modeling of Atmospheric Chemistry*. Cambridge University Press, Cambridge.

Brasseur, G.P., Xie, Y., Petersen, A.K., Bouarar, I., Flemming, J., Gauss, M., Jiang, F., Kouznetsov, R., Kranenburg, R., Mijling, B. et al. (2019). Ensemble forecasts of air quality in eastern China – Part 1: Model description and implementation of the MarcoPolo–Panda prediction system, version 1. *Geoscientific Model Development*, 12(1), 33–67.

Briant, R., Tuccella, P., Deroubaix, A., Khvorostyanov, D., Menut, L., Mailler, S., Turquety, S. (2017). Aerosol–radiation interaction modelling using online coupling between the WRF 3.7.1 meteorological model and the CHIMERE 2016 chemistry-transport model, through the OASIS3-MCT coupler. *Geoscientific Model Development*, 10(2), 927–944.

Brinkman, G.L., Milford, J.B., Schauer, J.J., Shafer, M.M., Hannigan, M.P. (2009). Source identification of personal exposure to fine particulate matter using organic tracers. *Atmospheric Environment*, 43(12), 1972–1981.

Brown, H., Liu, X., Pokhrel, R., Murphy, S., Lu, Z., Saleh, R., Mielonen, T., Kokkola, H., Bergman, T., Myhre, G. et al. (2021). Biomass burning aerosols in most climate models are too absorbing. *Nature Communications*, 277(12), 2041–1723.

Brutsaert, W. (1998). Land-surface water vapor and sensible heat flux: Spatial variability, homogeneity and measurements scales. *Water Resources Research*, 34(10), 2433–2442.

Buccolieri, R. and Hang, J. (2019). *Recent Advances in Urban Ventilation Assessment and Flow Modeling*. MDPI AG, Basel.

Burman, J., Jonsson, L., Rutgersson, A. (2019). On possibilities to estimate local concentration variations with CFD-LES in real urban environments. *Environmental Fluid Mechanics*, 19, 719–750.

Calbo, J., Pan, W., Webster, M., Prinn, R.G., McRae, G.J. (1998). Parameterization of urban subgrid scale processes in global atmospheric chemistry models. *Journal of Geophysical Research: Atmospheres*, 103(D3), 3437–3451.

Campbell, P., Zhang, Y., Yahya, K., Wang, K., Hogrefe, C., Pouliot, G., Knote, C., Hodzic, A., Jose, R.S., Perez, J.L. et al. (2015). A multi-model assessment for the 2006 and 2010 simulations under the Air Quality Model Evaluation International Initiative (AQMEII) phase 2 over North America. Part I: Indicators of the sensitivity of O3 and PM2.5 formation regimes. *Atmospheric Environment*, 115, 569–586.

Capps, S.L., Henze, D.K., Hakami, A., Russell, A.G., Nenes, A. (2012). Anisorropia: The adjoint of the aerosol thermodynamic model isorropia. *Atmospheric Chemistry and Physics*, 12(1), 527–543.

Carslaw, K.S., Boucher, O., Spracklen, D.V., Mann, G.W., Rae, J.G.L., Woodward, S., Kulmala, M. (2010). A review of natural aerosol interactions and feedbacks within the earth system. *Atmospheric Chemistry and Physics*, 10(4), 1701–1737.

Carter, W. (2010). Development of the SAPRC-07 chemical mechanism. *Atmospheric Environment*, 44(40), 5324–5335.

Cecil, D.J., Buechler, D.E., Blakeslee, R.J. (2014). Gridded lightning climatology from TRMM-LIS and OTD: Dataset description. *Atmospheric Research*, 135–136, 404–414.

Chang, J. and Hanna, S. (2004). Air quality model performance evaluation. *Meteorology and Atmospheric Physics*, 87, 167–196.

Chang, M., Hartley, D., Chang, W.-L. (1996). Inverse modeling of biogenic isoprene emissions. *Geophysical Research Letters*, 23(21), 3007.

Chang, M., Hartley, D., Cardelino, C., Haas-Laursen, D., Chang, W. (1997). On using inverse methods for resolving emissions with large spatial inhomogeneities. *Journal of Geophysical Research*, 102(D13), 16023–16036.

Chapman, E.G., Gustafson Jr., W.I., Easter, R.C., Barnard, J.C., Ghan, S.J., Pekour, M.S., Fast, J.D. (2009). Coupling aerosol-cloud-radiative processes in the WRF-Chem model: Investigating the radiative impact of elevated point sources. *Atmospheric Chemistry and Physics*, 9(3), 945–964.

Chatenet, B., Marticorena, B., Gomes, L., Bergametti, G. (1996). Assessing the microped size distributions of desert soils erodible by wind. *Sedimentology*, 43, 901–911.

Chatzidiakou, L., Krause, A., Popoola, O.A.M., Di Antonio, A., Kellaway, M., Han, Y., Squires, F.A., Wang, T., Zhang, H., Wang, Q. et al. (2019). Characterising low-cost sensors in highly portable platforms to quantify personal exposure in diverse environments. *Atmospheric Measurement Techniques*, 12(8), 4643–4657.

Chazette, P., Dabas, A., Sanak, J., Lardier, M., Royer, P. (2012). French airborne lidar measurements for Eyjafjallajökull ash plume survey. *Atmospheric Chemistry and Physics*, 12(15), 7059–7072.

Cholakian, A., Beekmann, M., Coll, I., Ciarelli, G., Colette, A. (2019). Biogenic secondary organic aerosol sensitivity to organic aerosol simulation schemes in climate projections. *Atmospheric Chemistry and Physics*, 19(20), 13209–13226.

Choobari, O.A., Zawar-Reza, P., Sturman, A. (2014). The global distribution of mineral dust and its impacts on the climate system: A review. *Atmospheric Research*, 138, 152–165.

Claquin, T., Schulz, M., Balkanski, Y.J. (1999). Modeling the mineralogy of atmospheric dust sources. *Journal of Geophysical Research: Atmospheres*, 104(D18), 22243–22256.

Clark, S.K., Ward, D.S., Mahowald, N.M. (2017). Parameterization-based uncertainty in future lightning flash density. *Geophysical Research Letters*, 44(6), 2893–2901.

Clerbaux, C., Hadji-Lazaro, J., Hauglustaine, D., Mégie, G., Khattatov, B., Lamarque, J.-F. (2001). Assimilation of carbon monoxide measured from satellite in a three-dimensional chemistry-transport model. *Journal of Geophysical Research: Atmospheres*, 106(D14), 15385–15394.

Clerbaux, C., Boynard, A., Clarisse, L., George, M., Hadji-Lazaro, J., Herbin, H., Hurtmans, D., Pommier, M., Razavi, A., Turquety, S. et al. (2009). Monitoring of atmospheric composition using the thermal infrared IASI/MetOp sounder. *Atmospheric Chemistry and Physics*, 9(16), 6041–6054.

Cofała, J., Amann, M., Klimont, Z., Kupiainen, K., Höglund-Isaksson, L. (2007). Scenarios of global anthropogenic emissions of air pollutants and methane until 2030. *Atmospheric Environment*, 41(38), 8486–8499.

Colella, P. and Woodward, P.R. (1984). The piecewise parabolic method (PPM) for gas-dynamical simulations. *Journal of Computational Physics*, 11, 38–39.

Colette, A., Favez, O., Meleux, F., Chiappini, L., Haeffelin, M., Morille, Y., Malherbe, L., Papin, A., Bessagnet, B., Menut, L. et al. (2011). Assessing in near real time the impact of the April 2010 Eyjafjallajökull ash plume on air quality. *Atmospheric Environment*, 45, 1217–1221.

Colette, A., Granier, C., Hodnebrog, Ø., Jakobs, H., Maurizi, A., Nyiri, A., Rao, S., Amann, M., Bessagnet, B., D'Angiola, A. et al. (2012). Future air quality in Europe: A multi-model assessment of projected exposure to ozone. *Atmospheric Chemistry and Physics*, 12(21), 10613–10630.

Colette, A., Bessagnet, B., Vautard, R., Szopa, S., Rao, S., Schucht, S., Klimont, Z., Menut, L., Clain, G., Meleux, F. et al. (2013). European atmosphere in 2050, a regional air quality and climate perspective under CMIP5 scenarios. *Atmospheric Chemistry and Physics*, 13, 7451–7471.

Constantinescu, E.M., Sandu, A., Chai, T., Carmichael, G.R. (2007). Ensemble-based chemical data assimilation I: General approach. *Quarterly Journal of the Royal Meteorological Society*, 133(626), 1229–1243.

Courant, R., Issacson, E., Reees, M. (1952). On the solution of nonlinear hyperbolic differential equations by finite differences. *Communications on Pure and Applied Mathematics*, 5, 243–255.

Couvidat, F. and Sartelet, K. (2015). The secondary organic aerosol processor (SOAP v1.0) model: A unified model with different ranges of complexity based on the molecular surrogate approach. *Geoscientific Model Development*, 8(4), 1111–1138.

Cox, W.M. and Tikvart, J.A. (1990). A statistical procedure for determining the best performing air quality simulation model. *Atmospheric Environment. Part A. General Topics*, 24(9), 2387–2395.

Crippa, M., Solazzo, E., Huang, G., Guizzardi, D., Koffi, E., Muntean, M., Schieberle, C., Friedrich, R., Janssens-Maenhout, G. (2020). High resolution temporal profiles in the emissions database for global atmospheric research. *Nature-Scientific Data*, 7(121), 17.

Cros, B., Durand, P., Cachier, H., Drobinski, P., Fréjafon, E., Kottmeir, C., Perros, P., Peuch, V.-H., Ponche, J.-L., Robin, D. et al. (2004). The ESCOMPTE program: An overview. *Atmospheric Research*, 69, 241–279.

Dabberdt, W.F. and Miller, E. (2000). Uncertainty, ensembles and air quality dispersion modeling: Applications and challenges. *Atmospheric Environment*, 34, 4667–4673.

Darbah, J., Kubiske, M., Nelson, N., Oksanen, E., Vapaavuori, E. (2008). Effects of decadal exposure to interacting elevated CO2 and/or O3 on paper birch (*Betula papyrifera*) reproduction. *Environmental Pollution*, 155, 446–452.

D'Arrigo, R., Seager, R., Smerdon, J., LeGrande, A., Cook, E.R. (2011). The anomalous winter of 1783-1784: Was the Laki eruption or an analog of the 2009–2010 winter to blame? *Geophysical Research Letters*, 38, L05706.

Davoine, X. and Bocquet, M. (2007). Inverse modelling-based reconstruction of the Chernobyl source term available for long-range transport. *Atmospheric Chemistry and Physics*, 7(6), 1549–1564.

De Ridder, K., Lefebre, F., Adriaensen, S., Arnold, U., Beckroege, W., Bronner, C., Damsgaard, O., Dostal, I., Dufek, J., Hirsch, J. et al. (2008). Simulating the impact of urban sprawl on air quality and population exposure in the German Ruhr area. Part ii: Development and evaluation of an urban growth scenario. *Atmospheric Environment*, 42(30), 7070–7077.

Deardoff, J. (1972). Numerical investigation of neutral and unstable planetary boundary layers. *Journal of the Atmospheric Sciences*, 29, 91–115.

DeCaria, A.J., Pickering, K.E., Stenchikov, G.L., Scala, J.R., Stith, J.L., Dye, J.E., Ridley, B.A., Laroche, P. (2000). A cloud-scale model study of lightning-generated NO_x in an individual thunderstorm during STERAO-A. *Journal of Geophysical Research: Atmospheres*, 105(D9), 11601–11616.

Delle Monache, L. and Stull, R.B. (2003). An ensemble air quality forecast over western Europe during an ozone episode. *Atmospheric Environment*, 37, 3469–3474.

Delle Monache, L., Deng, X., Zhou, Y., Stull, R. (2006). Ozone ensemble forecasts 1: A new ensemble design. *Journal of Geophysical Research*, 111, D05307.

Delmas, R., Mégie, G., Peuch, V. (2005). *Physique et chimie de l'atmosphère*. Éditions Belin, Paris.

DeMott, P.J., Prenni, A.J., Liu, X., Kreidenweis, S.M., Petters, M.D., Twohy, C.H., Richardson, M.S., Eidhammer, T., Rogers, D.C. (2010). Predicting global atmospheric ice nuclei distributions and their impacts on climate. *Proceedings of the National Academy of Sciences*, 107(25), 11217–11222.

Demuzere, M. and van Lipzig, N.P. (2010). A new method to estimate air-quality levels using a synoptic-regression approach. Part I: Present-day O3 and PM10 analysis. *Atmospheric Environment*, 44(10), 1341–1355.

Denby, B., Schaap, M., Segers, A., Builtjes, P., Horálek, J. (2008). Comparison of two data assimilation methods for assessing PM10 exceedances on the European scale. *Atmospheric Environment*, 42(30), 7122–7134.

Déqué, M., Rowell, D.P., Lüthi, D., Giorgi, F., Christensen, J.H., Rockel, B., Jacob, D., Kjellström, E., de Castro, M., van den Hurk, B. (2007). An intercomparison of regional climate simulations for Europe: Assessing uncertainties in model projections. *Climatic Change*, 81(70), 53.

Deschamps, P.-Y., Breon, F.-M., Leroy, M., Podaire, A., Bricaud, A., Buriez, J.-C., Seze, G. (1994). The POLDER mission: Instrument characteristics and scientific objectives. *Geoscience and Remote Sensing*, 32, 598–615.

Dipu, S., Quaas, J., Wolke, R., Stoll, J., Mühlbauer, A., Sourdeval, O., Salzmann, M., Heinold, B., Tegen, I. (2017). Implementation of aerosol–cloud interactions in the regional atmosphere–aerosol model COSMO-MUSCAT(5.0) and evaluation using satellite data. *Geoscientific Model Development*, 10(6), 2231–2246.

Donahue, N., Robinson, A., Stanier, C., Pandis, S.N. (2006). Coupled partitioning, dilution, and chemical aging of semivolatile organics. *Environmental Science and Technology*, 40(8), 2635–2643.

Dubovik, O., Smirnov, A., Holben, B.N., King, M.D., Kaufman, Y.J., Eck, T.F., Slutsker, I. (2000). Accuracy assessments of aerosol optical properties retrieved from Aerosol Robotic Network (AERONET) Sun and sky radiance measurements. *Journal of Geophysical Research*, 105(D8), 9791–9806.

Duhanyan, N. and Roustan, Y. (2011). Below-cloud scavenging by rain of atmospheric gases and particulates. *Atmospheric Environment*, 45(39), 7201–7217.

Dulac, F., Arboledas, L.A., Alastuey, A., Ancellet, G., Arndt, J., Attié, J.-L., Augustin, P., Becagli, S., Bergametti, G., Bocquet, M. et al. (2013). Overview of the project ChArMEx activities on Saharan dust in the Mediterranean region. In *7th International Workshop on Sand/Duststorms and Associated Dustfall*, 2–4 December, Frascati.

Dupont, E., Menut, L., Carissimo, B., Pelon, J., Flamant, P. (1999). Comparison between the atmospheric boundary layer in Paris and its rural suburbs during the ECLAP Experiment. *Atmospheric Environment*, 33, 979–994.

Eder, B., Bash, J., Foley, K., Pleim, J. (2014). Incorporating principal component analysis into air quality model evaluation. *Atmospheric Environment*, 82, 307–315.

Efstathiou, C., Isukapalli, S., Georgopoulos, P. (2011). A mechanistic modeling system for estimating large-scale emissions and transport of pollen and co-allergens. *Atmospheric Environment*, 45(13), 2260–2276.

Elbern, H. and Schmidt, H. (1999). A four-dimensional variational chemistry data assimilation scheme for Eulerian chemistry transport modeling. *Journal of Geophysical Research*, 104, 18583–18598.

Elbern, H., Schmidt, H., Ebel, A. (1997). Variational data assimilation for tropospheric chemistry modeling. *Journal of Geophysical Research*, 102, 15967–15985.

Elguindi, N., Granier, C., Stavrakou, T., Darras, S., Bauwens, M., Cao, H., Chen, C., Denier van der Gon, H.A.C., Dubovik, O., Fu, T.M. et al. (2020). Intercomparison of magnitudes and trends in anthropogenic surface emissions from bottom-up inventories, top-down estimates, and emission scenarios. *Earth's Future*, 8(8), e2020EF001520.

Elliott, S. (2009). Dependence of DMS global sea-air flux distribution on transfer velocity and concentration field type. *Journal of Geophysical Research: Biogeosciences*, 114(G2). doi: 10.1029/2008JG000710.

Emberson, L.-D., Simpson, D., Tuovinen, J.-P., Ashmore, M.-R., Cambridge, H.-M. (2000). Towards a model of ozone deposition and stomatal uptake over Europe. Technical Report, EMEP/MSC-W 2000.

Emeis, S., Forkel, R., Junkermann, W., Schäfer, K., Flentje, H., Gilge, S., Fricke, W., Wiegner, M., Freudenthaler, V., Groβ, S. et al. (2011). Measurement and simulation of the 16/17 April 2010 Eyjafjallajökull volcanic ash layer dispersion in the northern Alpine region. *Atmospheric Chemistry and Physics*, 11(6), 2689–2701.

Emmerson, K.M., Cope, M.E., Galbally, I.E., Lee, S., Nelson, P.F. (2018). Isoprene and monoterpene emissions in south-east Australia: Comparison of a multi-layer canopy model with MEGAN and with atmospheric observations. *Atmospheric Chemistry and Physics*, 18(10), 7539–7556.

Emmons, L.K., Walters, S., Hess, P.G., Lamarque, J.-F., Pfister, G.G., Fillmore, D., Granier, C., Guenther, A., Kinnison, D., Laepple, T. et al. (2010). Description and evaluation of the model for ozone and related chemical tracers, version 4 (MOZART-4). *Geoscientific Model Development*, 3(1), 43–67.

Engelstaedter, S., Tegen, I., Washington, R. (2006). North African dust emissions and transport. *Earth-Science Reviews*, 79(1), 73–100.

Enting, I. (2002). *Inverse Problems in Atmospheric Constituent Transport*. Cambridge University Press, Cambridge.

Erisman, J.W., van Pul, A., Wyers, P. (1994). Parameterization of surface resistance for the quantification of atmospheric deposition of acidifying pollutants and ozone. *Atmospheric Environment*, 28, 2595–2607.

Errico, R.M. (1997). What is an adjoint model? *Bulletin of the American Meteorological Society*, 78(11), 2577–2592.

Falasca, S. and Curci, G. (2018). High-resolution air quality modeling: Sensitivity tests to horizontal resolution and urban canopy with WRF-CHIMERE. *Atmospheric Environment*, 187, 241–254.

Fang, Y., Fiore, A.M., Horowitz, L.W., Gnanadesikan, A., Held, I., Chen, G., Vecchi, G., Levy, H. (2011). The impacts of changing transport and precipitation on pollutant distributions in a future climate. *Journal of Geophysical Research: Atmospheres*, 116(D18). doi: 10.1029/2011JD015642.

Fecan, F., Marticorena, B., Bergametti, G. (1999). Parameterization of the increase of aeolian erosion threshold wind friction velocity due to soil moisture for arid and semi-arid areas. *Annals of Geophysics*, 17, 149–157.

Feng, L., Smith, S.J., Braun, C., Crippa, M., Gidden, M.J., Hoesly, R., Klimont, Z., van Marle, M., van den Berg, M., van der Werf, G.R. (2020). The generation of gridded emissions data for CMIP6. *Geoscientific Model Development*, 13(2), 461–482.

Fenger, J. (2009). Air pollution in the last 50 years – From local to global. *Atmospheric Environment*, 43, 13–22.

Fesquet, C., Drobinski, P., Barthlott, C., Dubos, T. (2009). Impact of terrain heterogeneity on near-surface turbulence structure. *Atmospheric Research*, 94, 254–269.

Finn, D., Clawson, K.L., Carter, R.G., Rich, J.D., Biltoft, C., Leach, M. (2010). Analysis of urban atmosphere plume concentration fluctuations. *Boundary-Layer Meteorology*, 136, 431–456.

Finney, D.L., Doherty, R.M., Wild, O., Huntrieser, H., Pumphrey, H.C., Blyth, A.M. (2014). Using cloud ice flux to parametrise large-scale lightning. *Atmospheric Chemistry and Physics*, 14(23), 12665–12682.

Finney, D.L., Doherty, R.M., Wild, O., Abraham, N.L. (2016). The impact of lightning on tropospheric ozone chemistry using a new global lightning parametrisation. *Atmospheric Chemistry and Physics*, 16(12), 7507–7522.

Fishbain, B., Lerner, U., Castell, N., Cole-Hunter, T., Popoola, O., Broday, D.M., Iniguez, T.M., Nieuwenhuijsen, M., Jovasevic-Stojanovic, M., Topalovic, D. et al. (2017). An evaluation tool kit of air quality micro-sensing units. *Science of the Total Environment*, 575, 639–648.

Flagg, D.D. and Taylor, P.A. (2011). Sensitivity of mesoscale model urban boundary layer meteorology to the scale of urban representation. *Atmospheric Chemistry and Physics*, 11(6), 2951–2972.

Forkel, R., Werhahn, J., Hansen, A.B., McKeen, S., Peckham, S., Grell, G., Suppan, P. (2012). Effect of aerosol-radiation feedback on regional air quality – A case study with WRF/Chem. *Atmospheric Environment*, 53, 202–211.

Fountoukis, C. and Nenes, A. (2007). Isorropia II: A computationally efficient thermodynamic equilibrium model for K^+-Ca^{2+}-Mg^{2+}-$Nh_4{}^+$-Na^+-$SO_4{}^{2-}$-$NO_3{}^-$-Cl^--H_2O aerosols. *Atmospheric Chemistry and Physics*, 7(17), 4639–4659.

Fowler, D., Pilegaard, K., Sutton, M., Ambus, P., Raivonen, M., Duyzer, J., Simpson, D., Fagerli, H., Fuzzi, S., Schjoerring, J. et al. (2009). Atmospheric composition change: Ecosystems-atmosphere interactions. *Atmospheric Environment*, 43(33), 5193–5267.

Frei, C., Schöll, R., Fukutome, S., Schmidli, J., Vidale, P. (2006). Future change of precipitation extremes in Europe: Intercomparison of scenarios from regional climate models. *Journal of Geophysical Research*, 111, D06105.

Freitas, S.R., Longo, K.M., Chatfield, R., Latham, D., Silva Dias, M.A.F., Andreae, M.O., Prins, E., Santos, J.C., Gielow, R., Carvalho Jr., J.A. (2007). Including the sub-grid scale plume rise of vegetation fires in low resolution atmospheric transport models. *Atmospheric Chemistry and Physics*, 7, 3385–3398.

Freitas, S.R., Panetta, J., Longo, K.M., Rodrigues, L.F., Moreira, D.S., Rosário, N.E., Silva Dias, P.L., Silva Dias, M.A.F., Souza, E.P., Freitas, E.D. et al. (2017). The Brazilian developments on the Regional Atmospheric Modeling System (BRAMS 5.2): An integrated environmental model tuned for tropical areas. *Geoscientific Model Development*, 10(1), 189–222.

Fuzzi, S., Baltensperger, U., Carslaw, K., Decesari, S., Denier van der Gon, H., Facchini, M.C., Fowler, D., Koren, I., Langford, B., Lohmann, U. et al. (2015). Particulate matter, air quality and climate: Lessons learned and future needs. *Atmospheric Chemistry and Physics*, 15(14), 8217–8299.

Gaines Wilson, J. and Zawar-Reza, P. (2006). Intraurban-scale dispersion modelling of particulate matter concentrations: Applications for exposure estimates in cohort studies. *Atmospheric Environment*, 40(6), 1053–1063.

Galmarini, S., Bianconi, R., Klug, W., Mikkelsen, T., Addis, R., Andronopoulos, S., Astrup, P., Baklanov, A., Bartniki, J., Bartzis, J. et al. (2004). Ensemble dispersion forecasting. Part I: Concept, approach and indicators. *Atmospheric Environment*, 38, 4607–4617.

Galmarini, S., Bianconi, R., Appel, W., Solazzo, E., Mosca, S., Grossi, P., Moran, M., Schere, K., Rao, S. (2012). ENSEMBLE and AMET: Two systems and approaches to a harmonized, simplified and efficient facility for air quality models development and evaluation. *Atmospheric Environment*, 53, 51–59.

Gantt, B. and Meskhidze, N. (2013). The physical and chemical characteristics of marine primary organic aerosol: A review. *Atmospheric Chemistry and Physics*, 13(8), 3979–3996.

Garaud, D. and Mallet, V. (2010). Automatic generation of large ensembles for air quality forecasting using the polyphemus system. *Geoscientific Model Development*, 3(1), 69–85.

Garcia-Menendez, F. and Odman, M.T. (2011). Adaptive grid use in air quality modeling. *Atmosphere*, 2(3), 484–509.

Georgoulias, A.K., van der A, R.J., Stammes, P., Boersma, K.F., Eskes, H.J. (2019). Trends and trend reversal detection in 2 decades of tropospheric NO_2 satellite observations. *Atmospheric Chemistry and Physics*, 19(9), 6269–6294.

Gidden, M.J., Riahi, K., Smith, S.J., Fujimori, S., Luderer, G., Kriegler, E., van Vuuren, D.P., van den Berg, M., Feng, L., Klein, D. et al. (2019). Global emissions pathways under different socioeconomic scenarios for use in CMIP6: A dataset of harmonized emissions trajectories through the end of the century. *Geoscientific Model Development*, 12(4), 1443–1475.

Giglio, L., van der Werf, G.R., Randerson, J.T., Collatz, G.J., Kasibhatla, P. (2006). Global estimation of burned area using MODIS active fire observations. *Atmospheric Chemistry and Physics*, 6(4), 957–974.

Giorgi, F. (1986). A particle dry deposition scheme for use in tracer transport models. *Journal of Geophysical Research: Atmospheres*, 91, 9794–9806.

Giorgi, F. (2019). Thirty years of regional climate modeling: Where are we and where are we going next? *Journal of Geophysical Research: Atmospheres*, 124(11), 5696–5723.

Gordillo-Vázquez, F., Pérez-Invernón, F., Huntrieser, H., Smith, A. (2019). Comparison of six lightning parameterizations in CAM5 and the impact on global atmospheric chemistry. *Earth and Space Science*, 6(12), 2317–2346.

Gordon, H., Field, P.R., Abel, S.J., Dalvi, M., Grosvenor, D.P., Hill, A.A., Johnson, B.T., Miltenberger, A.K., Yoshioka, M., Carslaw, K.S. (2018). Large simulated radiative effects of smoke in the south-east Atlantic. *Atmospheric Chemistry and Physics*, 18(20), 15261–15289.

Gorkowski, K., Preston, T.C., Zuend, A. (2019). Relative-humidity-dependent organic aerosol thermodynamics via an efficient reduced-complexity model. *Atmospheric Chemistry and Physics*, 19(21), 13383–13407.

Granier, C., Darras, S., van der Gon, H.D., Doubalova, J., Elguindi, N., Galle, B., Gauss, M., Guevara, M., Jalkanen, J.-P., Kuenen, J. et al. (2019). The Copernicus Atmosphere Monitoring Service global and regional emissions (April 2019 version). Copernicus Atmosphere Monitoring Service (CAMS) Report, 16, 54.

Gratsea, M., Bösch, T., Kokkalis, P., Richter, A., Vrekoussis, M., Kazadzis, S., Tsekeri, A., Papayannis, A., Mylonaki, M., Amiridis, V. et al. (2020). Retrieval and evaluation of tropospheric aerosol extinction profiles using MAX-DOAS measurements over Athens, Greece. *Atmospheric Measurement Techniques Discussions*, 2020, 1–30.

Grell, G.A. and Baklanov, A. (2011). Integrated modeling for forecasting weather and air quality: A call for fully coupled approaches. *Atmospheric Environment*, 45, 6845–6851.

Grell, G.A. and Freitas, S.R. (2014). A scale and aerosol aware stochastic convective parameterization for weather and air quality modeling. *Atmospheric Chemistry and Physics*, 14(10), 5233–5250.

Gressent, A., Sauvage, B., Cariolle, D., Evans, M., Leriche, M., Mari, C., Thouret, V. (2016). Modeling lightning-NO_x chemistry on a sub-grid scale in a global chemical transport model. *Atmospheric Chemistry and Physics*, 16(9), 5867–5889.

Grini, A., Myhre, G., Sundet, J.K., Isaksen, I.S.A. (2002). Modeling the annual cycle of sea salt in the global 3D model Oslo CTM2: Concentrations, fluxes, and radiative impact. *Journal of Climate*, 15(13), 1717–1730.

Grythe, H., Ström, J., Krejci, R., Quinn, P., Stohl, A. (2014). A review of sea-spray aerosol source functions using a large global set of sea salt aerosol concentration measurements. *Atmospheric Chemistry and Physics*, 14(3), 1277–1297.

Guelle, W., Balkanski, Y.J., Dibb, J.E., Schulz, M., Dulac, F. (1998). Wet deposition in a global size-dependent aerosol transport model 2. Influence of the scavenging scheme on Pb vertical profiles, and deposition. *Journal of Geophysical Research: Atmospheres*, 103(D22), 28875–28891.

Guenther, A., Karl, T., Harley, P., Wiedinmyer, C., Palmer, P.I., Geron, C. (2006). Estimates of global terrestrial isoprene emissions using MEGAN (Model of Emissions of Gases and Aerosols from Nature). *Atmospheric Chemistry and Physics*, 6(11), 3181–3210.

Guenther, A.B., Jiang, X., Heald, C.L., Sakulyanontvittaya, T., Duhl, T., Emmons, L.K., Wang, X. (2012). The model of emissions of gases and aerosols from nature version 2.1 (MEGAN2.1): An extended and updated framework for modeling biogenic emissions. *Geoscientific Model Development*, 5(6), 1471–1492.

Guevara, M., Tena, C., Porquet, M., Jorba, O., Pérez García-Pando, C. (2020). HERMESv3, a stand-alone multi-scale atmospheric emission modelling framework – Part 2: The bottom-up module. *Geoscientific Model Development*, 13(3), 873–903.

Gurney, K., Chen, Y.-H., Maki, T., Kawa, S., Andrews, A., Zhu, Z. (2005). Sensitivity of atmospheric CO_2 inversions to seasonal and interannual variations in fossil fuel emissions. *Journal of Geophysical Research*, 110. doi: 10.1029/2004JD005373.

Hakami, A., Odman, M.T., Russell, A.G. (2003). High-order, direct sensitivity analysis of multidimensional air quality models. *Environmental Science & Technology*, 37(11), 2442–2452.

Hakami, A., Odman, M.T., Russell, A.G. (2004). Nonlinearity in atmospheric response: A direct sensitivity analysis approach. *Journal of Geophysical Research: Atmospheres*, 109(D15). doi: 10.1029/2003JD004502.

Hamill, T.M., Mullen, S.L., Snyder, C., Toth, Z., Baumhefner, D.P. (2000). Ensemble forecasting in the short to medium range: Report from a workshop. *Bulletin of the American Meteorological Society*, 81, 2653–2664.

Han, Z., Li, J., Guo, W., Xiong, Z., Zhang, W. (2013). A study of dust radiative feedback on dust cycle and meteorology over East Asia by a coupled regional climate-chemistry-aerosol model. *Atmospheric Environment*, 68(0), 54–63.

Hanea, R.G., Velders, G.J.M., Heemink, A. (2004). Data assimilation of ground-level ozone in Europe with a Kalman filter and chemistry transport model. *Journal of Geophysical Research: Atmospheres*, 109(D10). doi: 10.1029/2003JD004283.

Harikishan, G., Padmakumari, B., Maheskumar, R., Kulkarni, J. (2015). Radiative effect of dust aerosols on cloud microphysics and meso-scale dynamics during monsoon breaks over Arabian sea. *Atmospheric Environment*, 105(0), 22–31.

Harrison, R.M., Dall'Osto, M., Beddows, D.C.S., Thorpe, A.J., Bloss, W.J., Allan, J.D., Coe, H., Dorsey, J.R., Gallagher, M., Martin, C. et al. (2012). Atmospheric chemistry and physics in the atmosphere of a developed megacity (London): An overview of the REPARTEE experiment and its conclusions. *Atmospheric Chemistry and Physics*, 12(6), 3065–3114.

Haywood, J. and Boucher, O. (2000). Estimates of the direct and indirect radiative forcing due to tropospheric aerosols: A review. *Reviews of Geophysics*, 38, 513–543.

He, S., Carmichael, G.R., Sandu, A., Hotchkiss, B., Damian-Iordache, V. (2000). Application of ADIFOR for air pollution model sensitivity studies. *Environmental Modelling and Software: With Environment Data News*, 15(6), 549–557.

Heimann, M. and Kaminski, T. (1999). Inverse modeling approaches to infer surface trace gas fluxes from observed atmospheric mixing ratios. *Developments in Atmospheric Science*, 24, 277–295.

Heinold, B., Tegen, I., Schepanski, K., Tesche, M., Esselborn, M., Freudenthaler, V., Gross, S., Kandler, K., Knippertz, P., Müller, D. et al. (2011). Regional modelling of Saharan dust and biomass-burning smoke. *Tellus B: Chemical and Physical Meteorology*, 63(4), 781–799.

Helbig, N., Vogel, B., Vogel, H., Fiedler, F. (2004). Numerical modelling of pollen dispersion on the regional scale. *Aerobiologia*, 3, 3–19.

Hennemuth, B. and Lammert, A. (2006). Determination of the atmospheric boundary layer height from radiosonde and lidar backscatter. *Boundary-Layer Meteorology*, 120(1), 181–200.

Henze, D.K., Hakami, A., Seinfeld, J.H. (2007). Development of the adjoint of GEOS-Chem. *Atmospheric Chemistry and Physics*, 7(9), 2413–2433.

Hertel, O., Berkowicz, R., Christensen, J., Hov, O. (1993). Test of two numerical schemes for use in atmospheric transport-chemistry models. *Atmospheric Environment*, 27A(16), 2591–2611.

Hicks, B.B., Baldocchi, D.D., Meyers, T.P., Hosker Jr., R.P., Matt, D.R. (1987). A preliminary multiple resistance routine for deriving dry deposition velocities from measured quantities. *Water, Air, Soil & Pollution*, 36, 311–330.

Highwood, E. and Stevenson, D. (2003). Atmospheric impact of the 1783–1784 Laki eruption. Part II: Climatic effect of sulphate aerosol. *Atmospheric Chemistry and Physics*, 3, 1177–1189.

Hoesly, R.M., Smith, S.J., Feng, L., Klimont, Z., Janssens-Maenhout, G., Pitkanen, T., Seibert, J.J., Vu, L., Andres, R.J., Bolt, R.M. et al. (2018). Historical (1750–2014) anthropogenic emissions of reactive gases and aerosols from the Community Emissions Data System (CEDS). *Geoscientific Model Development*, 11(1), 369–408.

Holben, B., Eck, T., Slutsker, I., Tanré, D., Buis, J., Setzer, A., Vermote, E., Reagan, J., Kaufman, Y., Nakajima, T. et al. (1998). AERONET: A federated instrument network and data archive for aerosol characterization. *Remote Sensing of Environment*, 66, 1–16.

Höller, H., Betz, H.-D., Schmidt, K., Calheiros, R.V., May, P., Houngninou, E., Scialom, G. (2009). Lightning characteristics observed by a VLF/LF lightning detection network (LINET) in Brazil, Australia, Africa and Germany. *Atmospheric Chemistry and Physics*, 9(20), 7795–7824.

Holtslag, A., Van Meijgaard, E., De Rooy, W. (1995). A comparison of boundary layer diffusion schemes in unstable conditions over land. *Boundary-Layer Meteorology*, 76, 69–95.

Hong, S.Y., Noh, Y., Dudhia, J. (2006). A new vertical diffusion package with an explicit treatment of entrainment processes. *Monthly Weather Review*, 134, 2318–2341.

Honoré, C. and Vautard, R. (2000). Photochemical regimes in urban atmospheres: The influence of dispersion. *Geophysical Research Letters*, 27, 1895–1898.

Honoré, C., Rouïl, L., Vautard, R., Beekmann, M., Bessagnet, B., Dufour, A., Elichegaray, C., Flaud, J., Malherbe, L., Meleux, F. et al. (2008). Predictability of European air quality: The assessment of three years of operational forecasts and analyses by the PREV'AIR system. *Journal of Geophysical Research*, 113, D04301.

Houweling, S., Kaminski, T., Dentener, F., Lelieveld, J., Heinmann, M. (1999). Inverse modeling of methane sources and sinks using the adjoint of a global transport model. *Journal of Geophysical Research*, 104(D21), 26137–26160.

Huntrieser, H., Schumann, U., Schlager, H., Höller, H., Giez, A., Betz, H.-D., Brunner, D., Forster, C., Pinto Jr., O., Calheiros, R. (2008). Lightning activity in Brazilian thunderstorms during TROCCINOX: Implications for NO_x production. *Atmospheric Chemistry and Physics*, 8(4), 921–953.

Im, U. (2013). Impact of sea-salt emissions on the model performance and aerosol chemical composition and deposition in the East Mediterranean coastal regions. *Atmospheric Environment*, 75, 329–340.

Im, U., Brandt, J., Geels, C., Hansen, K.M., Christensen, J.H., Andersen, M.S., Solazzo, E., Kioutsioukis, I., Alyuz, U., Balzarini, A. et al. (2018a). Assessment and economic valuation of air pollution impacts on human health over Europe and the United States as calculated by a multi-model ensemble in the framework of AQMEII3. *Atmospheric Chemistry and Physics*, 18(8), 5967–5989.

Im, U., Christensen, J., Geels, C., Hansen, K., Brandt, J., Solazzo, E., Alyuz, U., Balzarini, A., Baro, R., Bellasio, R. et al. (2018b). Influence of anthropogenic emissions and boundary conditions on multi-model simulations of major air pollutants over Europe and North America in the framework of AQMEII3. *Atmospheric Chemistry and Physics*, 18(12), 8929–8952.

Jacob, D. (1999). *Introduction to Atmospheric Chemistry*. Princeton University Press, Princeton.

Jacob, D.J. and Winner, D.A. (2009). Effect of climate change on air quality. *Atmospheric Environment*, 43(1), 51–63.

Jacob, D.J., Crawford, J.H., Maring, H., Clarke, A.D., Dibb, J.E., Emmons, L.K., Ferrare, R.A., Hostetler, C.A., Russell, P.B., Singh, H.B. et al. (2010). The Arctic Research of the Composition of the Troposphere from Aircraft and Satellites (ARCTAS) mission: Design, execution, and first results. *Atmospheric Chemistry and Physics*, 10(11), 5191–5212.

Jacobson, M.Z. (1999). *Fundamentals of Atmospheric Modeling*. Cambridge University Press, Cambridge.

Jacobson, M.Z. and Streets, D.G. (2009). Influence of future anthropogenic emissions on climate, natural emissions, and air quality. *Journal of Geophysical Research: Atmospheres*, 114(D8). doi: 10.1029/2008JD011476.

Jang, J.-C., Jeffries, H., Byun, D., Pleim, J. (1995). Sensitivity of ozone to model grid resolution – I. Application of high-resolution regional acid deposition model. *Atmospheric Environment*, 29, 3085–3100.

Jégou, F., Godin-Beekman, S., Corrêa, M.P., Brogniez, C., Auriol, F., Peuch, V.H., Haeffelin, M., Pazmino, A., Saiag, P., Goutail, F. et al. (2011). Validity of satellite measurements used for the monitoring of UV radiation risk on health. *Atmospheric Chemistry and Physics*, 11(24), 13377–13394.

Jiang, J., Aksoyoglu, S., Ciarelli, G., Oikonomakis, E., El-Haddad, I., Canonaco, F., O'Dowd, C., Ovadnevaite, J., Minguillón, M.C., Baltensperger, U. et al. (2019). Effects of two different biogenic emission models on modelled ozone and aerosol concentrations in Europe. *Atmospheric Chemistry and Physics*, 19(6), 3747–3768.

Jiménez, P.A., Dudhia, J., González-Rouco, J.F., Navarro, J., Montávez, J.P., Garcia-Bustamante, E. (2012). A revised scheme for the WRF surface layer formulation. *Monthly Weather Review*, 140(3), 898–918.

Kaiser, J.W., Heil, A., Andreae, M.O., Benedetti, A., Chubarova, N., Jones, L., Morcrette, J.-J., Razinger, M., Schultz, M.G., Suttie, M. et al. (2012). Biomass burning emissions estimated with a global fire assimilation system based on observed fire radiative power. *Biogeosciences*, 9(1), 527–554.

Kalnay, E., Kanamitsu, M., Kistler, R., Collins, W., Deaven, D., Gandin, L., Iredell, M., Saha, S., White, G., Woollen, J. et al. (1996). The NCEP/NCAR 40-year reanalysis project. *Bulletin of the American Meteorological Society*, 77(3), 437–471.

Kaminski, T. and Heimann, M. (2001). Inverse modeling of atmospheric carbon dioxide fluxes. *Science*, 294, 259.

Kanakidou, M., Seinfeld, J.H., Pandis, S.N., Barnes, I., Dentener, F.J., Facchini, M.C., Van Dingenen, R., Ervens, B., Nenes, A., Nielsen, C.J. et al. (2005). Organic aerosol and global climate modelling: A review. *Atmospheric Chemistry and Physics*, 5(4), 1053–1123.

Kang, D., Foley, K.M., Mathur, R., Roselle, S.J., Pickering, K.E., Allen, D.J. (2019a). Simulating lightning NO production in CMAQV5.2: Performance evaluations. *Geoscientific Model Development*, 12(10), 4409–4424.

Kang, D., Pickering, K.E., Allen, D.J., Foley, K.M., Wong, D.C., Mathur, R., Roselle, S.J. (2019b). Simulating lightning NO production in CMAQV5.2: Evolution of scientific updates. *Geoscientific Model Development*, 12(7), 3071–3083.

Kang, D., Mathur, R., Pouliot, G., Gilliam, R., Wong, D. (2020). Significant ground-level ozone attributed to lightning-induced nitrogen oxides during summertime over the Mountain West States. *Nature Climate and Atmospheric Science*, 3(6), 7.

Karydis, V.A., Capps, S.L., Russell, A.G., Nenes, A. (2012). Adjoint sensitivity of global cloud droplet number to aerosol and dynamical parameters. *Atmospheric Chemistry and Physics*, 12(19), 9041–9055.

Kassomenos, P., Papaloukas, C., Petrakis, M., Karakitsios, S. (2008). Assessment and prediction of short term hospital admissions: The case of Athens, Greece. *Atmospheric Environment*, 42(30), 7078–7086.

Katragkou, E., Zanis, P., Kioutsioukis, I., Tegoulias, I., Melas, D., Kruger, B.C., Coppola, E. (2011). Future climate change impacts on summer surface ozone from regional climate-air quality simulations over Europe. *Journal of Geophysical Research: Atmospheres*, 116(D22). doi: 10.1029/2011JD015899.

Kaynak, B., Hu, Y., Martin, R.V., Russell, A.G., Choi, Y., Wang, Y. (2008). The effect of lightning NO_x production on surface ozone in the continental United States. *Atmospheric Chemistry and Physics*, 8(17), 5151–5159.

Kettle, A.J. and Andreae, M.O. (2000). Flux of dimethylsulfide from the oceans: A comparison of updated data sets and flux models. *Journal of Geophysical Research: Atmospheres*, 105(D22), 26793–26808.

Kim, K.-H., Kabir, E., Kabir, S. (2015). A review on the human health impact of airborne particulate matter. *Environment International*, 74, 136–143.

Kington, J. (1988). *The Weather of the 1780s Over Europe*. Cambridge University Press, Cambridge.

Kinne, S., Lohmann, U., Feichter, J., Schulz, M., Timmreck, C., Ghan, S., Easter, R., Chin, M., Ginoux, P., Takemura, T. et al. (2003). Monthly averages of aerosol properties: A global comparison among models, satellite data, and AERONET ground data. *Journal of Geophysical Research*, 108(D20), 4634.

Kioutsioukis, I. and Galmarini, S. (2014). De praeceptis ferendis: Good practice in multi-model ensembles. *Atmospheric Chemistry and Physics*, 14(21), 11791–11815.

Klimont, Z., Kupiainen, K., Heyes, C., Purohit, P., Cofala, J., Rafaj, P., Borken-Kleefeld, J., Schöpp, W. (2017). Global anthropogenic emissions of particulate matter including black carbon. *Atmospheric Chemistry and Physics*, 17(14), 8681–8723.

Knippertz, P., Coe, H., Chiu, J.C., Evans, M.J., Fink, A.H., Kalthoff, N., Liousse, C., Mari, C., Allan, R.P., Brooks, B. et al. (2015). The DACCIWA project: Dynamics-aerosol-chemistry-cloud interactions in West Africa. *Bulletin of the American Meteorological Society*, 96(9), 1451–1460.

Kok, J.F., Mahowald, N.M., Fratini, G., Gillies, J.A., Ishizuka, M., Leys, J.F., Mikami, M., Park, M.-S., Park, S.-U., Van Pelt, R.S. et al. (2014). An improved dust emission model Ž013. Part 1: Model description and comparison against measurements. *Atmospheric Chemistry and Physics*, 14(23), 13023–13041.

Kolehmainen, M., Martikainen, H., Ruuskanen, J. (2001). Neural networks and periodic components used in air quality forecasting. *Atmospheric Environment*, 35(5), 815–825.

Konovalov, I.B., Beekmann, M., Richter, A., Burrows, J. (2005). Inverse modelling of the spatial distribution of NO_x emissions on a continental scale using satellite data. *Atmospheric Chemistry and Physics*, 5, 12641–12695.

Konovalov, I.B., Beekmann, M., Burrows, J.P., Richter, A. (2008). Satellite measurement based estimates of decadal changes in European nitrogen oxides emissions. *Atmospheric Chemistry and Physics*, 8(10), 2623–2641.

Konovalov, I.B., Beekmann, M., Berezin, E.V., Formenti, P., Andreae, M.O. (2017). Probing into the aging dynamics of biomass burning aerosol by using satellite measurements of aerosol optical depth and carbon monoxide. *Atmospheric Chemistry and Physics*, 17(7), 4513–4537.

Kopacz, M., Jacob, D.J., Fisher, J.A., Logan, J.A., Zhang, L., Megretskaia, I.A., Yantosca, R.M., Singh, K., Henze, D.K., Burrows, J.P. et al. (2010). Global estimates of CO sources with high resolution by adjoint inversion of multiple satellite datasets (MOPITT, AIRS, SCIAMACHY, TES). *Atmospheric Chemistry and Physics*, 10(3), 855–876.

Kot, S. and Song, Y. (1998). An improvement of the Louis scheme for the surface layer in an atmospheric modelling system. *Boundary-Layer Meteorology*, 88, 239–254.

Krpo, A., Salamanca, F., Martilli, A., Clappier, A. (2010). On the impact of anthropogenic heat fluxes on the urban boundary layer: A two-dimensional numerical study. *Boundary-Layer Meteorology*, 136(1), 105–127.

Krysta, M., Bocquet, M., Brandt, J. (2008). Probing ETEX-II data set with inverse modelling. *Atmospheric Chemistry and Physics*, 8(14), 3963–3971.

Kulmala, M., Laaksonen, A., Pirjola, L. (1998). Parameterization for sulfuric acid/water nucleation rates. *Journal of Geophysical Research: Atmospheres*, 103(No D7), 8301–8307.

Kulmala, M., Asmi, A., Lappalainen, H.K., Baltensperger, U., Brenguier, J.-L., Facchini, M.C., Hansson, H.-C., Hov, Ø., O'Dowd, C.D., Pöschl, U. et al. (2011). General overview: European integrated project on aerosol cloud climate and air quality interactions (EUCAARI) – Integrating aerosol research from nano to global scales. *Atmospheric Chemistry and Physics*, 11(24), 13061–13143.

Lac, C., Chaboureau, J.-P., Masson, V., Pinty, J.-P., Tulet, P., Escobar, J., Leriche, M., Barthe, C., Aouizerats, B., Augros, C. et al. (2018). Overview of the Meso-NH model version 5.4 and its applications. *Geoscientific Model Development*, 11(5), 1929–1969.

Laj, P., Klausen, J., Bilde, M., Plas-Duelmer, C., Pappalardo, G., Clerbaux, C., Baltensperger, U., Hjorth, J., Simpson, D., Reimann, S. et al. (2009). Measuring atmospheric composition change. *Atmospheric Environment*, 43(33), 5351–5414.

Lana, A., Bell, T.G., Simo, R., Vallina, S.M., Ballabrera-Poy, J., Kettle, A.J., Dachs, J., Bopp, L., Saltzman, E.S., Stefels, J. et al. (2011). An updated climatology of surface dimethlysulfide concentrations and emission fluxes in the global ocean. *Global Biogeochemical Cycles*, 25(1). doi: 10.1029/2010GB003850.

Lattuati, M. (1997). Contribution à l'étude du bilan de l'ozone troposphérique à l'interface de l'Europe et de l'Atlantique Nord : modélisation lagrangienne et mesures en altitude. Phd Thesis, Pierre and Marie Curie University, Paris.

Lebel, T., Parker, D.J., Flamant, C., Bourlès, B., Marticorena, B., Mougin, E., Peugeot, C., Diedhiou, A., Haywood, J.M., Ngamini, J.B. et al. (2010). The AMMA field campaigns: Multiscale and multidisciplinary observations in the West African region. *Quarterly Journal of the Royal Meteorological Society*, 136(S1), 8–33.

Leibensperger, E.M., Mickley, L.J., Jacob, D.J., Chen, W.-T., Seinfeld, J.H., Nenes, A., Adams, P.J., Streets, D.G., Kumar, N., Rind, D. (2012). Climatic effects of 1950–2050 changes in US anthropogenic aerosols – Part 1: Aerosol trends and radiative forcing. *Atmospheric Chemistry and Physics*, 12(7), 3333–3348.

Lemonsu, A. and Masson, V. (2002). Simulation of a summer urban breeze over Paris. *Boundary-Layer Meteorology*, 104, 463–490.

Lettau, H. (1969). Note on aerodynamic roughness-parameter estimation on the basis of roughness-element description. *Journal of Applied Meteorology*, 8, 828–832.

Leutbecher, M. and Palmer, T.N. (2008). Ensemble forecasting. *Journal of Computational Physics*, 227(7), 3515–3539.

Li, G., Wang, Y., Zhang, R. (2008). Implementation of a two-moment bulk microphysics scheme to the WRF model to investigate aerosol-cloud interaction. *Journal of Geophysical Research: Atmospheres*, 113(D15). doi: 10.1029/2007JD009361.

Li, Y.P., Elbern, H., Lu, K.D., Friese, E., Kiendler-Scharr, A., Mentel, T.F., Wang, X.S., Wahner, A., Zhang, Y.H. (2013). Updated aerosol module and its application to simulate secondary organic aerosols during impact campaign May 2008. *Atmospheric Chemistry and Physics*, 13(13), 6289–6304.

Lightstone, S.D., Moshary, F., Gross, B. (2017). Comparing CMAQ forecasts with a neural network forecast model for PM2.5 in New York. *Atmosphere*, 8(9), 161. doi: 10.3390/atmos8090161.

Lim, K.-S.S., Fan, J., Leung, L.R., Ma, P.-L., Singh, B., Zhao, C., Zhang, Y., Zhang, G., Song, X. (2013). Investigation of aerosol indirect effects using a cumulus microphysics parameterization in a regional climate model. *Journal of Geophysical Research: Atmospheres*, 119(2), 906–926.

Liss, P.S. and Merlivat, L. (1986). *Air-Sea Gas Exchange Rates: Introduction and Synthesis*. Springer, Dordrecht.

Löndahl, J., Swietlicki, E., Lindgren, E., Loft, S. (2010). Aerosol exposure versus aerosol cooling of climate: What is the optimal emission reduction strategy for human health? *Atmospheric Chemistry and Physics*, 10(19), 9441–9449.

Loosmore, G. and Cederwall, R. (2004). Precipitation scavenging of atmospheric aerosols for emergency response applications: Testing an updated model with new real-time data. *Atmospheric Environment*, 38, 993–1003.

Lorenz, E.N. (1963). Deterministic non-periodic flow. *Journal of the Atmospheric Sciences*, 20, 130–141.

Lou, S., Russell, L.M., Yang, Y., Liu, Y., Singh, B., Ghan, S.J. (2017). Impacts of interactive dust and its direct radiative forcing on interannual variations of temperature and precipitation in winter over East Asia. *Journal of Geophysical Research: Atmospheres*, 2017JD027267.

Louis, J., Tiedke, M., Geleyn, J. (1982). A short history of the PBL parametrization at ECMWF. In *ECMWF Workshop on Planetary Boundary Layer Parametrization*, University of Reading, Reading.

Loveland, T., Reed, B., Brown, J., Ohlen, D., Zhu, J., Yang, L., Merchant, J. (2000). Development of a global land cover characteristics database and IGBP discover from 1-km AVHRR data. *International Journal of Remote Sensing*, 21(6/7), 1303–1330.

Lundgren, K., Vogel, B., Vogel, H., Kottmeier, C. (2013). Direct radiative effects of sea salt for the Mediterranean region under conditions of low to moderate wind speeds. *Journal of Geophysical Research: Atmospheres*, 118(4), 1906–1923.

Luo, C., Wang, Y., Koshak, W.J. (2017). Development of a self-consistent lightning NO_x simulation in large-scale 3-D models. *Journal of Geophysical Research: Atmospheres*, 2016JD026225.

Madronich, S. and Calvert, J.G. (1990). The NCAR master mechanism of the gas phase chemistry version 2.0. NCAR Technical Notes, NCAR/TN-333+SRT.

Mahowald, N., Prinn, R., Rasch, P. (1997). Deducing CCL3 F emissions using an inverse method and chemical transport models with assimilated winds. *Journal of Geophysical Research*, 102, 28153–28168.

Mahowald, N., Ward, D.S., Kloster, S., Flanner, M.G., Heald, C.L., Heavens, N.G., Hess, P.G., Lamarque, J.-F., Chuang, P.Y. (2011). Aerosol impacts on climate and biogeochemistry. *Annual Review of Environment and Resources*, 36(1), 45–74.

Mahrt, L. and Vickers, D. (2004). Bulk formulation of the surface heat flux. *Boundary-Layer Meteorology*, 110, 357–379.

Mailler, S., Menut, L., Khvorostyanov, D., Valari, M., Couvidat, F., Siour, G., Turquety, S., Briant, R., Tuccella, P., Bessagnet, B. et al. (2017). CHIMERE-2017: From urban to hemispheric chemistry-transport modeling. *Geoscientific Model Development*, 10(6), 2397–2423.

Makra, L., Juhasz, M., Borsos, E., Beczi, R. (2004). Meteorological variables connected with airborne ragweed pollen in Southern Hungary. *International Journal of Biometeorology*, 49, 37–47.

Malmqvist, E., Oudin, A., Pascal, M., Medina, S. (2018). Choices behind numbers: A review of the major air pollution health impact assessments in Europe. *Current Environmental Health Reports*, 5, 34.

Marchand, M., Hilburn, K., Miller, S.D. (2019). Geostationary lightning mapper and earth networks lightning detection over the contiguous United States and dependence on flash characteristics. *Journal of Geophysical Research: Atmospheres*. doi: 10.1029/2019JD031039.

Marécal, V., Peuch, V.-H., Andersson, C., Andersson, S., Arteta, J., Beekmann, M., Benedictow, A., Bergström, R., Bessagnet, B., Cansado, A. et al. (2015). A regional air quality forecasting system over Europe: The MACC-II daily ensemble production. *Geoscientific Model Development*, 8(9), 2777–2813.

Mari, C., Jacob, D.J., Bechtold, P. (2000). Transport and scavenging of soluble gases in a deep convective cloud. *Journal of Geophysical Research: Atmospheres*, 105(D17), 22255–22267.

Mari, C., Chaboureau, J.P., Pinty, J.P., Duron, J., Mascart, P., Cammas, J.P., Gheusi, F., Fehr, T., Schlager, H., Roiger, A. et al. (2006). Regional lightning NO_x sources during the TROCCINOX experiment. *Atmospheric Chemistry and Physics*, 6(12), 5559–5572.

Martensson, E.M., Nilsson, E.D., de Leeuw, G., Cohen, L.H., Hansson, H.-C. (2003). Laboratory simulations and parameterization of the primary marine aerosol production. *Journal of Geophysical Research: Atmospheres*, 108(D9). doi: 10.1029/2002JD002263.

Marticorena, B. and Bergametti, G. (1995). Modeling the atmospheric dust cycle 1: Design of a soil derived dust production scheme. *Journal of Geophysical Research*, 100, 16415–16430.

Martilli, A., Clappier, A., Rotach, M. (2002). An urban surface exchange parameterisation for mesoscale models. *Boundary-Layer Meteorology*, 104, 261–304.

Martin, R.V. (2008). Satellite remote sensing of surface air quality. *Atmospheric Environment*, 42(34), 7823–7843.

Masson, V. (2000). A physically-based scheme for the urban energy budget in atmospheric models. *Boundary-Layer Meteorology*, 94, 357–397.

McHenry, J.N., Ryan, W.F., Seaman, N.L., Coats, C.J., Pudykiewicz, J., Arunachalam, S., Vukovich, J.M. (2004). A real-time Eulerian photo-chemical model forecast system: Overview and initial ozone forecast performance in the northeast U.S. corridor. *Bulletin of the American Meteorological Society*, 85, 525–548.

McNorton, J., Wilson, C., Gloor, M., Parker, R.J., Boesch, H., Feng, W., Hossaini, R., Chipperfield, M.P. (2018). Attribution of recent increases in atmospheric methane through 3-D inverse modelling. *Atmospheric Chemistry and Physics*, 18(24), 18149–18168.

McRae, G., Goodin, W., Seinfeld, J. (1982). Development of a second generation mathematical model for urban air pollution I: Model formulation. *Atmospheric Environment*, 16, 679–696.

de Meij, A., Thunis, P., Bessagnet, B., Cuvelier, C. (2009). The sensitivity of the CHIMERE model to emissions reduction scenarios on air quality in northern Italy. *Atmospheric Environment*, 43(11), 1897–1907.

Meijer, E., van Velthoven, P., Brunner, D., Huntrieser, H., Kelder, H. (2001). Improvement and evaluation of the parameterisation of nitrogen oxide production by lightning. *Physics and Chemistry of the Earth, Part C: Solar, Terrestrial and Planetary Science*, 26(8), 577–583.

Meleux, F., Solmon, F., Giorgi, F. (2007). Increase in European summer ozone amounts due to climate change. *Atmospheric Environment*, 41, 7577–7587.

Mellor, G.L. and Yamada, T. (1974). A hierarchy of turbulence closure models for planetary boundary layers. *Journal of the Atmospheric Sciences*, 31, 1791–1806.

Mendoza-Dominguez, A. and Russell, A. (2001). Estimation of emission adjustments from the application of four-dimensional data assimilation to photochemical air quality modeling. *Atmospheric Environment*, 35, 2879–2894.

Menut, L. (2003). Adjoint modelling for atmospheric pollution processes sensitivity at regional scale during the ESQUIF IOP2. *Journal of Geophysical Research*, 108(D17), 8562.

Menut, L. (2018). Modeling of mineral dust emissions with a Weibull wind speed distribution including subgrid-scale orography variance. *Journal of Atmospheric and Oceanic Technology*, 35(6), 1221–1236.

Menut, L. and Bessagnet, B. (2010). Atmospheric composition forecasting in Europe. *Annales Geophysicae*, 28(1), 61–74.

Menut, L. and Bessagnet, B. (2019). What can we expect from data assimilation for air quality forecast? Part I: Quantification with academic test cases. *Journal of Atmospheric and Oceanic Technology*, 36(2), 269–279.

Menut, L., Flamant, C., Pelon, J., Flamant, P. (1999). Urban boundary layer height determination from lidar measurements over the Paris area. *Applied Optics*, 38, 945–954.

Menut, L., Vautard, R., Flamant, C., Abonnel, C., Beekmann, M., Chazette, P., Flamant, P., Gombert, D., Guédalia, D., Lefebvre, M. et al. (2000). Measurements and modelling of atmospheric pollution over the Paris area: An overview of the ESQUIF Project. *Annales Geophysicae*, 18(11), 1467–1481.

Menut, L., Goussebaile, A., Bessagnet, B., Khvorostiyanov, D., Ung, A. (2012). Impact of realistic hourly emissions profiles on modelled air pollutants concentrations. *Atmospheric Environment*, 49, 233–244.

Menut, L., Perez Garcia-Pando, C., Haustein, K., Bessagnet, B., Prigent, C., Alfaro, S. (2013a). Relative impact of roughness and soil texture on mineral dust emission fluxes modeling. *Journal of Geophysical Research*, 118, 6505–6520.

Menut, L., Tripathi, O., Colette, A., Vautard, R., Flaounas, E., Bessagnet, B. (2013b). Evaluation of regional climate simulations for air quality modelling purposes. *Climate Dynamics*, 40, 2515–2533.

Menut, L., Mailler, S., Siour, G., Bessagnet, B., Turquety, S., Rea, G., Briant, R., Mallet, M., Sciare, J., Formenti, P., Meleux, F. (2015a). Ozone and aerosol tropospheric concentrations variability analyzed using the ADRIMED measurements and the WRF and CHIMERE models. *Atmospheric Chemistry and Physics*, 15(11), 6159–6182.

Menut, L., Rea, G., Mailler, S., Khvorostyanov, D., Turquety, S. (2015b). Aerosol forecast over the Mediterranean area during July 2013 (ADRIMED/CHARMEX). *Atmospheric Chemistry and Physics*, 15(14), 7897–7911.

Menut, L., Siour, G., Mailler, S., Couvidat, F., Bessagnet, B. (2016). Observations and regional modeling of aerosol optical properties, speciation and size distribution over Northern Africa and western Europe. *Atmospheric Chemistry and Physics*, 16(20), 12961–12982.

Menut, L., Mailler, S., Bessagnet, B., Siour, G., Colette, A., Couvidat, F., Meleux, F. (2017). An alternative way to evaluate chemistry-transport model variability. *Geoscientific Model Development*, 10(3), 1199–1208.

Menut, L., Flamant, C., Turquety, S., Deroubaix, A., Chazette, P., Meynadier, R. (2018). Impact of biomass burning on pollutant surface concentrations in megacities of the Gulf of Guinea. *Atmospheric Chemistry and Physics*, 18(4), 2687–2707.

Menut, L., Tuccella, P., Flamant, C., Deroubaix, A., Gaetani, M. (2019). The role of aerosol–radiation–cloud interactions in linking anthropogenic pollution over southern west Africa and dust emission over the Sahara. *Atmospheric Chemistry and Physics*, 19(23), 14657–14676.

Menut, L., Bessagnet, B., Mailler, S., Pennel, R., Siour, G. (2020a). Impact of lightning NO_x emissions on atmospheric composition and meteorology in Africa and Europe. *Atmosphere*, 11(10), 1128. doi: 10.3390/atmos11101128.

Menut, L., Siour, G., Bessagnet, B., Couvidat, F., Journet, E., Balkanski, Y., Desboeufs, K. (2020b). Modelling the mineralogical composition and solubility of mineral dust in the Mediterranean area with CHIMERE 2017r4. *Geoscientific Model Development*, 13(4), 2051–2071.

Menut, L., Bessagnet, B., Briant, R., Cholakian, A., Couvidat, F., Mailler, S., Pennel, R., Siour, G., Tuccella, P., Turquety, S., Valari, M. (2021). The CHIMERE v2020r1 online chemistry-transport model. *Geoscientific Model Development*, 14(11), 6781–6811.

Mestayer, P.G., Durand, P., Augustin, P., Bastin, S., Bonnefond, J.M., Benech, B., Campistron, B., Coppalle, A., Delbarre, H., Dousset, B. et al. (2005). The urban boundary-layer field campaign in Marseille (UBL/CLU-ESCOMPTE): Set-up and first results. *Boundary-Layer Meteorology*, 114, 315–365.

Metzger, S., Steil, B., Xu, L., Penner, J.E., Lelieveld, J. (2012). New representation of water activity based on a single solute specific constant to parameterize the hygroscopic growth of aerosols in atmospheric models. *Atmospheric Chemistry and Physics*, 12(12), 5429–5446.

Middleton, N. (2017). Desert dust hazards: A global review. *Aeolian Research*, 24, 53–63.

Milliez, M. and Carissimo, B. (2008). Computational fluid dynamical modelling of concentration fluctuations in an idealized urban area. *Boundary-Layer Meteorology*, 127, 241–259.

Miyazaki, K., Eskes, H.J., Sudo, K., Zhang, C. (2014). Global lightning NO_x production estimated by an assimilation of multiple satellite data sets. *Atmospheric Chemistry and Physics*, 14(7), 3277–3305.

Molina, L.T., Madronich, S., Gaffney, J.S., Apel, E., de Foy, B., Fast, J., Ferrare, R., Herndon, S., Jimenez, J.L., Lamb, B. et al. (2010). An overview of the MILAGRO 2006 campaign: Mexico city emissions and their transport and transformation. *Atmospheric Chemistry and Physics*, 10(18), 8697–8760.

Moltchanov, S., Levy, I., Etzion, Y., Lerner, U., Broday, D.M., Fishbain, B. (2015). On the feasibility of measuring urban air pollution by wireless distributed sensor networks. *Science of the Total Environment*, 502, 537–547.

Monahan, E.C. (1986). The ocean as a source of atmospheric particles. In *The Role of Air-Sea Exchange in Geochemical Cycling*, Buat-Ménard, P. (ed.). Kluwer Academic Publishers, Dordrecht.

Monks, P.S., Granier, C., Fuzzi, S., Stohl, A., Williams, M., Akimoto, H., Amann, M., Baklanov, A., Baltensperger, U., Bey, I. et al. (2009). Atmospheric composition change – Global and regional air quality. *Atmospheric Environment*, 43(33), 5268–5350.

Monks, P.S., Archibald, A.T., Colette, A., Cooper, O., Coyle, M., Derwent, R., Fowler, D., Granier, C., Law, K.S., Mills, G.E. et al. (2015). Tropospheric ozone and its precursors from the urban to the global scale from air quality to short-lived climate forcer. *Atmospheric Chemistry and Physics*, 15(15), 8889–8973.

Morcrette, J.-J., Boucher, O., Jones, L., Salmond, D., Bechtold, P., Beljaars, A., Benedetti, A., Bonet, A., Kaiser, J.W., Razinger, M. et al. (2009). Aerosol analysis and forecast in the European centre for medium-range weather forecasts integrated forecast system: Forward modeling. *Journal of Geophysical Research: Atmospheres*, 114(D6). doi: 10.1029/2008JD011235.

Morille, Y., Haeffelin, M., Drobinski, P., Pelon, J. (2007). Strat: An automated algorithm to retrieve the vertical structure of the atmosphere from single channel lidar data. *Journal of Atmospheric and Oceanic Technology*, 24(5), 761–775.

Morman, S.A. and Plumlee, G.S. (2013). The role of airborne mineral dusts in human disease. *Aeolian Research*, 9, 203–212.

Moss, R.H., Edmonds, J.A., Hibbard, K.A., Manning, M.R., Rose, S.K., van Vuuren, D.P., Carter, T.R., Emori, S., Kainuma, M., Kram, T. et al. (2010). The next generation of scenarios for climate change research and assessment. *Nature*, 463(1476-4687), 756.

Mulholland, M. and Seinfeld, J. (1995). Inverse air pollution modelling of urban-scale carbon monoxide emissions. *Atmos. Environ.*, 29, 497–516.

Müller, J.-F. and Stavrakou, T. (2005). Inversion of CO and NO_x emissions using the adjoint of the IMAGES model. *Atmos. Chem. Phys.*, 5, 1157–1186.

Murray, L.T. (2016). Lightning NO_x and impacts on air quality. *Current Pollution Reports*, 2(2), 115–133.

Nenes, A., Pilinis, C., Pandis, S. (1998). ISORROPIA: A new thermodynamic model for inorganic multicomponent atmospheric aerosols. *Aquatic Geochem.*, 4, 123–152.

Nester, K. and Panitz, H.-J. (2006). Sensitivity analysis by the adjoint chemistry transport model DRAIS for an episode in the Berlin Ozone (BERLIOZ) experiment. *Atmospheric Chemistry and Physics*, 6(8), 2091–2106.

Nickovic, S., Cvetkovic, B., Madonna, F., Rosoldi, M., Pejanovic, G., Petkovic, S., Nikolic, J. (2016). Cloud ice caused by atmospheric mineral dust – Part 1: Parameterization of ice nuclei concentration in the NMME-DREAM model. *Atmospheric Chemistry and Physics*, 16(17), 11367–11378.

Nightingale, P.D., Malin, G., Law, C.S., Watson, A.J., Liss, P.S., Liddicoat, M.I., Boutin, J., Upstill-Goddard, R.C. (2000). In situ evaluation of air-sea gas exchange parameterizations using novel conservative and volatile tracers. *Global Biogeochemical Cycles*, 14(1), 373–387.

Nikulin, G., Kjellstrom, E., Hansson, U., Jones, C., Strandberg, G., Ullerstig, A. (2011). Evaluation and future projections of temperature, precipitation and wind extremes over Europe in an ensemble of regional climate simulations. *Tellus*, 63A, 41–55.

Oke, T.R. (1982). The energetic basis of the urban heat island. *Quarterly Journal of the Royal Meteorological Society*, 108(455), 1–24.

Ott, L.E., Pickering, K.E., Stenchikov, G.L., Allen, D.J., DeCaria, A.J., Ridley, B., Lin, R.-F., Lang, S., Tao, W.-K. (2010). Production of lightning NO_x and its vertical distribution calculated from three-dimensional cloud-scale chemical transport model simulations. *Journal of Geophysical Research: Atmospheres*, 115(D4). doi: 10.1029/2009JD011880.

Pagowski, M., Grell, G.A., Devenyi, D., Peckham, S.E., McKeen, S.A., Gong, W., Monache, L.D., McHenry, J.N., McQueen, J., Lee, P. (2006). Application of dynamic linear regression to improve the skill of ensemble-based deterministic ozone forecasts. *Atmospheric Environment*, 40, 3240–3250.

Pan, X., Ichoku, C., Chin, M., Bian, H., Darmenov, A., Colarco, P., Ellison, L., Kucsera, T., da Silva, A., Wang, J. et al. (2020). Six global biomass burning emission datasets: Intercomparison and application in one global aerosol model. *Atmospheric Chemistry and Physics*, 20(2), 969–994.

Pandis, S., Wexler, A., Seinfeld, J. (1993). Secondary organic aerosol formation and transport II: Predicting the ambient secondary organic aerosol size distribution. *Atmospheric Environment*, 27A, 2403–2416.

Pappalardo, G., Mona, L., D'Amico, G., Wandinger, U., Adam, M., Amodeo, A., Ansmann, A., Apituley, A., Alados Arboledas, L., Balis, D. et al. (2013). Four-dimensional distribution of the 2010 Eyjafjallajökull volcanic cloud over Europe observed by EARLINET. *Atmospheric Chemistry and Physics*, 13(8), 4429–4450.

Pappalardo, G., Amodeo, A., Apituley, A., Comeron, A., Freudenthaler, V., Linné, H., Ansmann, A., Bösenberg, J., D'Amico, G., Mattis, I. et al. (2014). Earlinet: Towards an advanced sustainable European aerosol lidar network. *Atmospheric Measurement Techniques*, 7(8), 2389–2409.

Partridge, D.G., Vrugt, J.A., Tunved, P., Ekman, A.M.L., Struthers, H., Sorooshian, A. (2012). Inverse modelling of cloud-aerosol interactions – Part 2: Sensitivity tests on liquid phase clouds using a Markov chain Monte Carlo based simulation approach. *Atmospheric Chemistry and Physics*, 12(6), 2823–2847.

Paugam, R., Wooster, M., Freitas, S., Val Martin, M. (2016). A review of approaches to estimate wildfire plume injection height within large-scale atmospheric chemical transport models. *Atmospheric Chemistry and Physics*, 16(2), 907–925.

Paulson, C.A. (1970). The mathematical representation of wind speed and temperature profiles in the unstable atmospheric surface layer. *Journal of Applied Meteorology and Climatology*, 9(6), 857–861.

Penner, J.E., Zhou, C., Garnier, A., Mitchell, D.L. (2018). Anthropogenic aerosol indirect effects in cirrus clouds. *Journal of Geophysical Research: Atmospheres*, 123(20), 11652–11677.

Péré, J., Mallet, M., Pont, V., Bessagnet, B. (2010). Evaluation of an aerosol optical scheme in the chemistry-transport model CHIMERE. *Atmospheric Environment*, 44, 3688–3699.

Pereira, G., Siqueira, R., Rosário, N.E., Longo, K.L., Freitas, S.R., Cardozo, F.S., Kaiser, J.W., Wooster, M.J. (2016). Assessment of fire emission inventories during the South AMerican Biomass Burning Analysis (SAMBBA) experiment. *Atmospheric Chemistry and Physics*, 16(11), 6961–6975.

Pérez, C., Nickovic, S., Pejanovic, G., Baldasano, J., Ozsoy, E. (2006). Interactive dust-radiation modelling: A step to improve weather forecasts. *Journal of Geophysical Research*, 111, D16206.

Peterson, H.S. and Beasley, W.H. (2011). Possible catalytic effects of ice particles on the production of NO_x by lightning discharges. *Atmospheric Chemistry and Physics*, 11(19), 10259–10268.

Pétron, G., Granier, C., Khattatov, B., Yudin, V., Lamarque, J.-F., Emmons, L., Gille, J., Edwards, D. (2004). Monthly CO surface sources inventory based on the 2000–2001 MOPITT satellite data. *Geophysical Research Letters*, 31(21). doi: 10.1029/2004GL020560.

Pichelli, E., Ferretti, R., Cacciani, M., Siani, A.M., Ciardini, V., Di Iorio, T. (2014). The role of urban boundary layer investigated with high-resolution models and ground-based observations in Rome area: A step towards understanding parameterization potentialities. *Atmospheric Measurement Techniques*, 7(1), 315–332.

Pickering, K.E., Wang, Y., Tao, W.-K., Price, C., Müller, J.-F. (1998). Vertical distributions of lightning NO_x for use in regional and global chemical transport models. *Journal of Geophysical Research: Atmospheres*, 103(D23), 31203–31216.

Pielke, R. (2013). *Mesoscale Meteorological Modeling*. Academic Press, Cambridge, MA.

Pierce, T., Geron, C., Bender, L., Dennis, R., Tonnesen, G., Guenther, A. (1998). Influence of increased isoprene emissions on regional ozone modeling. *Journal of Geophysical Research: Atmospheres*, 103(D19), 25611–25629.

Pierce, R.B., Schaack, T., Al-Saadi, J.A., Fairlie, T.D., Kittaka, C., Lingenfelser, G., Natarajan, M., Olson, J., Soja, A., Zapotocny, T. et al. (2007). Chemical data assimilation estimates of continental U.S. ozone and nitrogen budgets during the intercontinental chemical transport experiment – North America. *Journal of Geophysical Research: Atmospheres*, 112(D12), n/a–n/a. D12S21.

Piironen, A.K. and Eloranta, E.W. (1995). Convective boundary layer mean depths and cloud geometrical properties obtained from volume imaging lidar data. *Journal of Geophysical Research*, 100, 25569–25576.

Pison, I., Menut, L., Blond, N. (2006). Inverse modeling of emissions for local photo-oxidant pollution: Testing a new methodology with kriging constraints. *Annales Geophysicae*, 24, 1523–1535.

Pison, I., Menut, L., Bergametti, G. (2007). Inverse modeling of surface NO_x anthropogenic emissions fluxes in the Paris area during the ESQUIF campaign. *Journal of Geophysical Research, Atmospheres*, 112, D24302.

Pleim, J.E. (2007). A combined local and nonlocal closure model for the atmospheric boundary layer. Part II: Application and evaluation in a mesoscale meteorological model. *Journal of Applied Meteorology and Climatology*, 46(9), 1396–1409.

Pleim, J.E. and Ran, L. (2011). Surface flux modeling for air quality applications. *Atmosphere*, 2(3), 271–302.

Porter, P.S., Rao, S.T., Hogrefe, C., Gego, E., Mathur, R. (2015). Methods for reducing biases and errors in regional photochemical model outputs for use in emission reduction and exposure assessments. *Atmospheric Environment*, 112, 178–188.

Powers, J.G., Klemp, J.B., Skamarock, W.C., Davis, C.A., Dudhia, J., Gill, D.O., Coen, J.L., Gochis, D.J., Ahmadov, R., Peckham, S.E. et al. (2017). The weather research and forecasting model: Overview, system efforts, and future directions. *Bulletin of the American Meteorological Society*, 98(8), 1717–1737.

Prank, M., Chapman, D., Bullock, J., Belmonte, J., Berger, U., Dahl, A., Jager, S., Kovtunenko, I., Magyar, D., Niemela, S. et al. (2013). An operational model for forecasting ragweed pollen release and dispersion in Europe. *Agricultural and Forest Meteorology*, 182, 43–53.

Prank, M., Sofiev, M., Tsyro, S., Hendriks, C., Semeena, V., Vazhappilly Francis, X., Butler, T., Denier van der Gon, H., Friedrich, R., Hendricks, J. et al. (2016). Evaluation of the performance of four chemical transport models in predicting the aerosol chemical composition in Europe in 2005. *Atmospheric Chemistry and Physics*, 16(10), 6041–6070.

Price, C. (2013). Lightning applications in weather and climate research. *Surveys in Geophysics*, 34(6), 755–767.

Price, C. and Rind, D. (1992). A simple lightning parameterization for calculating global lightning distributions. *Journal of Geophysical Research*, 97, 9919–9933.

Price, C. and Rind, D. (1993). What determines the cloud-to-ground lightning fraction in thunderstorms? *Geophysical Research Letters*, 20, 463–466.

Price, C. and Rind, D. (1994). Modeling global lightning distributions in a general circulation model. *Monthly Weather Review*, 122(8), 1930–1939.

Price, C., Penner, J., Prather, M. (1997). NO_x from lightning 1: Global distribution based on lightning physics. *Journal of Geophysical Research: Atmospheres*, 102(D5), 5929–5941.

Prigent, C., Jiménez, C., Catherinot, J. (2012). Comparison of satellite microwave backscattering (ASCAT) and visible/near-infrared reflectances (PARASOL) for the estimation of aeolian aerodynamic roughness length in arid and semi-arid regions. *Atmospheric Measurement Techniques*, 5(11), 2703–2712.

Pryor, S.C., Schoof, J.T., Barthelmie, R.J. (2005). Empirical downscaling of wind speed probability distributions. *Journal of Geophysical Research: Atmospheres*, 110(D19), D19109.

Pun, B.K. and Seigneur, C. (2007). Investigative modeling of new pathways for secondary organic aerosol formation. *Atmospheric Chemistry and Physics*, 7(9), 2199–2216.

Pye, H.O.T., Murphy, B.N., Xu, L., Ng, N.L., Carlton, A.G., Guo, H., Weber, R., Vasilakos, P., Appel, K.W., Budisulistiorini, S.H. et al. (2017). On the implications of aerosol liquid water and phase separation for organic aerosol mass. *Atmospheric Chemistry and Physics*, 17(1), 343–369.

Quélo, D., Mallet, V., Sportisse, B. (2005). Inverse modeling of NO_x emissions at regional scale over Northern France. Preliminary investigation of the second-order sensitivity. *Journal of Geophysical Research*, 110, D24310.

Rabier, F., Courtier, P., Talagrand, O. (1992). An application of adjoint models to sensitivity analysis. *Contributions to Atmospheric Physics*, 65, 177–192.

Raffuse, S., Craig, K., Larkin, N., Strand, T., Sullivan, D., Wheeler, N., Solomon, R. (2012). An evaluation of modeled plume injection height with satellite-derived observed plume height. *Atmosphere*, 3, 103–123.

Ravi, S., D'Odorico, P., Breshears, D.D., Field, J.P., Goudie, A.S., Huxman, T.E., Li, J., Okin, G.S., Swap, R.J., Thomas, A.D. et al. (2011). Aeolian processes and the biosphere. *Reviews of Geophysics*, 49(3), RG3001.

Rea, G., Turquety, S., Menut, L., Briant, R., Mailler, S., Siour, G. (2015). Source contributions to 2012 summertime aerosols in the Euro-Mediterranean region. *Atmospheric Chemistry and Physics*, 15(14), 8013–8036.

Reeves, C., Cunnold, D., Derwent, R., Dlugodencky, E., Edouard, S., Granier, C., Ménard, R., Novelli, P., Parrish, D. (2004). Determination of emissions from observations of atmospheric compounds. In *Emissions of Atmospheric Trace Compounds*, Granier, C., Artaxo, P., Reeves, C.E. (eds). Kluwer Academic Publishers, Amsterdam.

Reid, J.S., Koppmann, R., Eck, T., Eleuterio, D. (2005a). A review of biomass burning emissions. Part II: Intensive physical properties of biomass burning particles. *Atmospheric Chemistry and Physics*, 5(3), 799–825.

Reid, J.S., Eck, T.F., Christopher, S.A., Koppmann, R., Dubovik, O., Eleuterio, D.P., Holben, B.N., Reid, E.A., Zhang, J. (2005b). A review of biomass burning emissions. Part III: Intensive optical properties of biomass burning particles. *Atmospheric Chemistry and Physics*, 5(3), 827–849.

Reis, S., Seto, E., Northcross, A., Quinn, N.W., Convertino, M., Jones, R.L., Maier, H.R., Schlink, U., Steinle, S., Vieno, M. et al. (2015). Integrating modelling and smart sensors for environmental and human health. *Environmental Modelling and Software*, 74, 238–246.

Rieger, D., Bangert, M., Bischoff-Gauss, I., Förstner, J., Lundgren, K., Reinert, D., Schröter, J., Vogel, H., Zängl, G., Ruhnke, R. et al. (2015). ICON-ART 1.0 – A new online-coupled model system from the global to regional scale. *Geoscientific Model Development*, 8(6), 1659–1676.

Riemer, N., Ault, A.P., West, M., Craig, R.L., Curtis, J.H. (2019). Aerosol mixing state: Measurements, modeling, and impacts. *Reviews of Geophysics*, 57(2), 187–249.

Rio, C., Hourdin, F., Chédin, A. (2010). Numerical simulation of tropospheric injection of biomass burning products by pyro-thermal plumes. *Atmospheric Chemistry and Physics*, 10, 3463–3478.

Robeson, S.M. and Steyn, D.G. (1990). Evaluation and comparison of statistical forecast models for daily maximum ozone concentrations. *Atmospheric Environment*, 24, 303–312.

Robinson, A., Donahue, N., Shrivastava, M.K., Weitkamp, E.A., Sage, A.M., Grieshop, A.P., Lane, T.E., Pierce, J.R., Pandis, S.N. (2007). Rethinking organic aerosols: Semivolatile emissions and photochemical aging. *Science*, 315(5816), 1259–1262.

Rogers, C., Wayne, P., Macklin, E., Muilenberg, M., Wagner, C. (2006). Interaction of the onset of spring and elevated atmospheric CO2 on ragweed (*Ambrosia artemisiifolia* L.) pollen production. *Environmental Health Perspectives*, 114, 865–869.

Russo, A., Raischel, F., Lind, P.G. (2013). Air quality prediction using optimal neural networks with stochastic variables. *Atmospheric Environment*, 79, 822–830.

Saide, P.E., Kim, J., Song, C.H., Choi, M., Cheng, Y., Carmichael, G.R. (2014). Assimilation of next generation geostationary aerosol optical depth retrievals to improve air quality simulations. *Geophysical Research Letters*, 41(24), 9188–9196.

Sandu, A. and Chai, T. (2011). Chemical data assimilation: An overview. *Atmosphere*, 2(3), 426–463.

Sandu, A., Daescu, D., Carmichael, G. (2003). Direct and adjoint sensitivity analysis of chemical kinetic systems with KPP: Part I – Theory and software tools. *Atmospheric Environment*, 37, 5083–5096.

Sarrat, C., Lemonsu, A., Masson, V., Guedalia, D. (2006). Impact of urban heat island on regional atmospheric pollution. *Atmospheric Environment*, 40(10), 1743–1758.

Sarwar, G., Luecken, D., Yarwood, G., Whitten, G.Z., Carter, W.P.L. (2008). Impact of an updated carbon bond mechanism on predictions from the CMAQ modeling system: Preliminary assessment. *Journal of Applied Meteorology and Climatology*, 47(1), 3–14.

Saunders, S.M., Jenkin, M.E., Derwent, R.G., Pilling, M.J. (2003). Protocol for the development of the master chemical mechanism, MCM v3 (Part A): Tropospheric degradation of non-aromatic volatile organic compounds. *Atmospheric Chemistry and Physics*, 3(1), 161–180.

Saunier, O., Mathieu, A., Didier, D., Tombette, M., Quélo, D., Winiarek, V., Bocquet, M. (2013). An inverse modeling method to assess the source term of the Fukushima nuclear power plant accident using gamma dose rate observations. *Atmospheric Chemistry and Physics*, 13(22), 11403–11421.

Schaap, M., Vautard, R., Bergstrom, R., van Loon, M., Bessagnet, B., Brandt, J., Christensen, H., Cuvelier, K., Foltescu, V., Graff, A. et al. (2007). Evaluation of long-term aerosol simulations from seven air quality models and their ensemble in the EURODELTA study. *Atmospheric Environment*, 41, 2083–2097.

Schaap, M., Cuvelier, C., Hendriks, C., Bessagnet, B., Baldasano, J., Colette, A., Thunis, P., Karam, D., Fagerli, H., Graff, A. et al. (2015). Performance of European chemistry transport models as function of horizontal resolution. *Atmospheric Environment*, 112, 90–105.

Schell, B., Ackermann, I.J., Hass, H., Binkowski, F.S., Ebel, A. (2001). Modeling the formation of secondary organic aerosol within a comprehensive air quality model system. *Journal of Geophysical Research: Atmospheres*, 106(D22), 28275–28293.

Schepanski, K. (2018). Transport of mineral dust and its impact on climate. *Geosciences*, 8(5), 151. doi: 10.3390/geosciences8050151.

Schmidt, H. and Martin, D. (2003). Continental scale adjoint sensitivity of ozone concentrations in the Paris region during the ESQUIF episodes. *Journal of Geophysical Research*, 108(D17), 4313.

Schmidt, A., Ostro, B., Carslaw, K., Wilson, M., Throdarson, T., Mann, G., Simmons, A. (2011). Excess mortality in Europe following a future Laki-style Icelandic eruption. *PNAS*, 108, 15710–15715.

Schumann, U. and Huntrieser, H. (2007). The global lightning-induced nitrogen oxides source. *Atmospheric Chemistry and Physics*, 7(14), 3823–3907.

Seinfeld, J. and Pandis, S. (2006). *Atmospheric Chemistry and Physics: From Air Pollution to Climate Change*, 2nd edition. Wiley-Interscience, Hoboken, NJ.

Semeniuk, K. and Dastoor, A. (2020). Current state of atmospheric aerosol thermodynamics and mass transfer modeling: A review. *Atmosphere*, 11(2), 156. doi: 10.3390/atmos11020156.

Shao, Y. (2004). Physics and modelling of wind erosion. *Progress in Energy and Combustion Science*, 30, 501–543.

Shao, Y. and Lu, I. (2000). A simple expression for wind erosion threshold friction velocity. *Journal of Geophysical Research*, 105, 22437–22443.

Shao, Y., Wyrwoll, K.-H., Chappell, A., Huang, J., Lin, Z., McTainsh, G.H., Mikami, M., Tanaka, T.Y., Wang, X., Yoon, S. (2011). Dust cycle: An emerging core theme in earth system science. *Aeolian Research*, 2(4), 181–204.

Shi, Z., Huang, L., Li, J., Ying, Q., Zhang, H., Hu, J. (2020). Sensitivity analysis of the surface ozone and fine particulate matter to meteorological parameters in China. *Atmospheric Chemistry and Physics*, 20(21), 13455–13466.

Shin, H. and Hong, S.-Y. (2011). Intercomparison of planetary boundary-layer parametrizations in the WRF model for a single day from CASES-99. *Boundary-Layer Meteorology*, 139(2), 261–281.

Shrivastava, M., Fast, J., Easter, R., Gustafson Jr., W.I., Zaveri, R.A., Jimenez, J.L., Saide, P., Hodzic, A. (2011). Modeling organic aerosols in a megacity: Comparison of simple and complex representations of the volatility basis set approach. *Atmospheric Chemistry and Physics*, 11(13), 6639–6662.

Sicard, M., Guerrero-Rascado, J.L., Navas-Guzmán, F., Preißler, J., Molero, F., Tomás, S., Bravo-Aranda, J.A., Comerón, A., Rocadenbosch, F., Wagner, F. et al. (2012). Monitoring of the Eyjafjallajökull volcanic aerosol plume over the Iberian Peninsula by means of four EARLINET lidar stations. *Atmospheric Chemistry and Physics*, 12(6), 3115–3130.

Sillman, S. (1999). The relation between ozone, NO_x and hydrocarbons in urban and polluted rural environments. *Atmospheric Environment*, 33, 1821–1845.

Simon, H., Baker, K., Phillips, S. (2012). Compilation and interpretation of photochemical model performance statistics published between 2006 and 2012. *Atmospheric Environment*, 61, 124–139.

Simpson, D., Winiwarter, W., Borjesson, G., Cinderby, S., Ferreiro, A., Guenther, A., Hewitt, C.N., Janson, R., Khalil, M.A.K., Owen, S. et al. (1999). Inventorying emissions from nature in Europe. *Journal of Geophysical Research: Atmospheres*, 104(D7), 8113–8152.

Simpson, D., Fagerli, H., Jonson, J.-E., Tsyro, S., Wind, P., Tuovinen, J.-P. (2003). Transboundary acidification, eutrophication and ground level ozone in Europe. Part I: Unified EMEP model description. Technical Report, EMEP/Status Report, Norwegian Meteorological Institute.

Simpson, D., Benedictow, A., Berge, H., Bergström, R., Emberson, L.D., Fagerli, H., Flechard, C.R., Hayman, G.D., Gauss, M., Jonson, J.E. et al. (2012). The EMEP MSC-W chemical transport model – Technical description. *Atmospheric Chemistry and Physics*, 12(16), 7825–7865.

Sindelarova, K., Granier, C., Bouarar, I., Guenther, A., Tilmes, S., Stavrakou, T., Müller, J.-F., Kuhn, U., Stefani, P., Knorr, W. (2014). Global data set of biogenic VOC emissions calculated by the MEGAN model over the last 30 years. *Atmospheric Chemistry and Physics*, 14(17), 9317–9341.

Singal, S.P. (1993). Monitoring air pollution related meteorology using SODAR. *Applied Physics B: Photophysics and Laser Chemistry*, 57(65–82).

Sinyuk, A., Holben, B.N., Eck, T.F., Giles, D.M., Slutsker, I., Korkin, S., Schafer, J.S., Smirnov, A., Sorokin, M., Lyapustin, A. (2020). The AERONET Version 3 aerosol retrieval algorithm, associated uncertainties and comparisons to Version 2. *Atmospheric Measurement Techniques*, 13(6), 3375–3411.

Skamarock, W.C., Klemp, J.B., Duda, M.G., Fowler, L.D., Park, S.-H., Ringler, T.D. (2012). A multiscale nonhydrostatic atmospheric model using centroidal Voronoi tesselations and C-grid staggering. *Monthly Weather Review*, 140(9), 3090–3105.

Slinn, W. (1983). Precipitation scavenging. Atmospheric Sciences and Power Production 1979. Division of Biomedical Environmental Research, US Department of Energy, Washington, DC.

Snider, G., Weagle, C.L., Martin, R.V., van Donkelaar, A., Conrad, K., Cunningham, D., Gordon, C., Zwicker, M., Akoshile, C., Artaxo, P. et al. (2015). SPARTAN: A global network to evaluate and enhance satellite-based estimates of ground-level particulate matter for global health applications. *Atmospheric Measurement Techniques*, 8(1), 505–521.

Sofiev, M., Siljamo, P., Ranta, H., Rantio-Lehtimaki, A. (2006). Towards numerical forecasting of long-range air transport of birch pollen: Theoritical considerations and a feasability study. *International Journal of Biometeorology*, 50, 392–402.

Sofiev, M., Ermakova, T., Vankevich, R. (2012). Evaluation of the smoke-injection height from wild-land fires using remote-sensing data. *Atmospheric Chemistry and Physics*, 12(4), 1995–2006.

Sofiev, M., Siljamo, P., Ranta, H., Linkosalo, T., Jeager, S., Rasmussen, A., Rantio-Lehtimaki, A., Severova, E., Kukkonen, J. (2013a). A numerical model of birch pollen emission and dispersion in the atmosphere. Description of the emission module. *International Journal of Biometeorology*, 57, 45–58.

Sofiev, M., Vankevich, R., Ermakova, T., Hakkarainen, J. (2013b). Global mapping of maximum emission heights and resulting vertical profiles of wildfire emissions. *Atmospheric Chemistry and Physics*, 13(14), 7039–7052.

Solazzo, E. and Galmarini, S. (2015). Comparing apples with apples: Using spatially distributed time series of monitoring data for model evaluation. *Atmospheric Environment*, 112, 234–245.

Stajner, I., Winslow, N., Rood, R.B., Pawson, S. (2004). Monitoring of observation errors in the assimilation of satellite ozone data. *Journal of Geophysical Research: Atmospheres*, 109(D6). doi: 10.1029/2003JD004118.

Stensrud, D. (2011). *Parameterization Schemes: Keys to Understanding Numerical Weather Prediction Models*. Cambridge University Press, Cambridge.

Stevenson, D.S., Johnson, C.E., Highwood, E.J., Gauci, V., Collins, W.J., Derwent, R.G. (2003). Atmospheric impact of the 1783–1784 Laki eruption. Part I: Chemistry modelling. *Atmospheric Chemistry and Physics*, 3(3), 487–507.

Stockwell, W.R., Kirchner, F., Kuhn, M., Seefeld, S. (1997). A new mechanism for regional atmospheric chemistry modeling. *Journal of Geophysical Research: Atmospheres*, 102(D22), 25847–25879.

Stockwell, W.R., Lawson, C.V., Saunders, E., Goliff, W.S. (2012). A review of tropospheric atmospheric chemistry and gas-phase chemical mechanisms for air quality modeling. *Atmosphere*, 3(1), 1–32.

von Storch, H. and Zwiers, F. (2001). *Statistical Analysis in Climate Research*. Cambridge University Press, Cambridge.

Stull, R. (1988). *An Introduction to Boundary Layer Meteorology*. Kluwer Academic Publishers Group, Dordrecht.

Talagrand, O. (1997). Assimilation of observations: An introduction. *Journal of the Meteorological Society of Japan*, 75(1B), 191–209.

Talagrand, O. and Courtier, P. (1987). Variational assimilation of meteorological observations with the adjoint vorticity equations. Part I – Theory. *Quaterly Journal of the Royal Meteorological Society*, 113, 1311–1328.

Talagrand, O., Vautard, R., Strauss, B. (1997). Evaluation of probabilistic prediction systems. In *Workshop on Predictability*, October 20–22. ECMWF, Shinfield Park, Reading.

Talagrand, O., Vautard, R., Strauss, B. (1999). Evaluation of probabilistic prediction systems. In *Proceedings of the ECMWF Workshop on Predicability*, 20–22 October. ECMWF, Shinfield Park, Reading.

Tarín-Carrasco, P., Morales-Suárez-Varela, M., Im, U., Brandt, J., Palacios-Peña, L., Jiménez-Guerrero, P. (2019). Isolating the climate change impacts on air-pollution-related-pathologies over central and southern Europe – A modelling approach on cases and costs. *Atmospheric Chemistry and Physics*, 19(14), 9385–9398.

Thompson, G. and Eidhammer, T. (2014). A study of aerosols impacts on clouds and precipitation development in a large winter cyclone. *Journal of the Atmospheric Sciences*, 71, 3636–3659.

Veira, A., Kloster, S., Wilkenskjeld, S., Remy, S. (2015). Fire emission heights in the climate system – Part 1: Global plume height patterns simulated by ECHAM6-HAM2. *Atmospheric Chemistry and Physics*, 15(13), 7155–7171.

Verhoelst, T., Compernolle, S., Pinardi, G., Lambert, J.-C., Eskes, H.J., Eichmann, K.-U., Fjæraa, A.M., Granville, J., Niemeijer, S., Cede, A. et al. (2020). Ground-based validation of the Copernicus Sentinel-5P TROPOMI NO_2 measurements with the NDACC ZSL-DOAS, MAX-DOAS and Pandonia global networks. *Atmospheric Measurement Techniques Discussions*, 14(1), 1–40.

Verwer, J. (1994). Gauss-Seidel iteration for stiff ODES from chemical kinetics. *SIAM Journal on Scientific Computing*, 15(5), 1243–1250.

Verwer, J. and Simpson, D. (1995). Explicit methods for stiff ODEs from atmospheric chemistry. *Applied Numerical Mathematics*, 18(1), 413–430.

Vestreng, V., Ntziachristos, L., Semb, A., Reis, S., Isaksen, I.S.A., Tarrasón, L. (2009). Evolution of NO_x emissions in Europe with focus on road transport control measures. *Atmospheric Chemistry and Physics*, 9(4), 1503–1520.

Viana, M., Querol, X., Alastuey, A., Ballester, F., Llop, S., Esplugues, A., Fernandez-Patier, R., Garcia dos Santos, S., Herce, M. (2008). Characterising exposure to PM aerosols for an epidemiological study. *Atmospheric Environment*, 42(7), 1552–1568.

Vira, J. and Sofiev, M. (2012). On variational data assimilation for estimating the model initial conditions and emission fluxes for short-term forecasting of SOx concentrations. *Atmospheric Environment*, 46, 318–328.

Vivanco, M., Bessagnet, B., Cuvelier, C., Theobald, M., Tsyro, S., Pirovano, G., Aulinger, A., Bieser, J., Calori, G., Ciarelli, G. et al. (2017). Joint analysis of deposition fluxes and atmospheric concentrations of inorganic nitrogen and sulphur compounds predicted by six chemistry transport models in the frame of the EURODELTAIII project. *Atmospheric Environment*, 151, 152–175.

Vivanco, M.G., Theobald, M.R., García-Gómez, H., Garrido, J.L., Prank, M., Aas, W., Adani, M., Alyuz, U., Andersson, C., Bellasio, R. et al. (2018). Modelled deposition of nitrogen and sulfur in Europe estimated by 14 air quality model-systems: Evaluation, effects of changes in emissions and implications for habitat protection. *Atmospheric Chemistry and Physics Discussions*, 2018, 1–35.

Vogel, B., Vogel, H., Bäumer, D., Bangert, M., Lundgren, K., Rinke, R., Stanelle, T. (2009). The comprehensive model system COSMO-ART: Radiative impact of aerosol on the state of the atmosphere on the regional scale. *Atmospheric Chemistry and Physics*, 9(22), 8661–8680.

Vogel, H., Förstner, J., Vogel, B., Hanisch, T., Mühr, B., Schättler, U., Schad, T. (2014). Time-lagged ensemble simulations of the dispersion of the Eyjafjallajökull plume over Europe with COSMO-ART. *Atmospheric Chemistry and Physics*, 14(15), 7837–7845.

Vuolo, M., Menut, L., Chepfer, H. (2009). Impact of transport schemes accuracy on modelled dust concentrations variability. *Journal of Atmospheric and Oceanic Technology*, 26, 1135–1143.

Wang, Y., DeSilva, A., Goldenbaum, G., Dickerson, R. (1998). Nitric oxide production by simulated lightning: Dependence on current, energy and pressure. *Journal of Geophysical Research: Atmospheres*, 103(D15), 19149–19159.

Wang, Q., Li, Z., Guo, J., Zhao, C., Cribb, M. (2018). The climate impact of aerosols on the lightning flash rate: Is it detectable from long-term measurements? *Atmospheric Chemistry and Physics*, 18(17), 12797–12816.

Wang, L., Tai, A.P.K., Tam, C.-Y., Sadiq, M., Wang, P., Cheung, K.K.W. (2020). Impacts of future land use and land cover change on mid-21st-century surface ozone air quality: Distinguishing between the biogeophysical and biogeochemical effects. *Atmospheric Chemistry and Physics*, 20(19), 11349–11369.

Wanninkhof, R. (1992). Relationship between wind speed and gas exchange over the ocean. *Journal of Geophysical Research: Oceans*, 97(C5), 7373–7382.

Waquet, F., Peers, F., Goloub, P., Ducos, F., Thieuleux, F., Derimian, Y., Riedi, J., Chami, M., Tanré, D. (2014). Retrieval of the Eyjafjallajökull volcanic aerosol optical and microphysical properties from POLDER/PARASOL measurements. *Atmospheric Chemistry and Physics*, 14(4), 1755–1768.

Weger, M., Heinold, B., Engler, C., Schumann, U., Seifert, A., Fößig, R., Voigt, C., Baars, H., Blahak, U., Borrmann, S. et al. (2018). The impact of mineral dust on cloud formation during the Saharan dust event in April 2014 over Europe. *Atmospheric Chemistry and Physics*, 18(23), 17545–17572.

Weill, A., Klapisz, C., Strauss, B., Baudin, F., Jaupart, C., Van Grunderbeeck, P., Goutorbe, J.P. (1979). Measuring heat flux and structure functions of temperature fluctuations with an acoustic Doppler sodar. *Journal of Applied Meteorology*, 19(2), 199–205.

Wesely, M. (1989). Parameterization of surface resistances to gaseous dry deposition in regional-scale numerical models. *Atmospheric Environment*, 23(23), 1293–1304.

White, B.R. (1986). Particle dynamics in two-phase flows. In *Encyclopedia of Fluid Mechanics*, Cheremisinoff, N.P. (ed.). Gulf Publishing, Houston, TX.

Wilks, D. (2006). *Statistical Methods in the Atmospheric Sciences*. Elsevier, Amsterdam.

Winker, D.M., Vaughan, M.A., Omar, A., Hu, Y., Powell, K.A., Liu, Z., Hunt, W.H., Young, S.A. (2009). Overview of the CALIPSO mission and CALIOP data processing algorithms. *Journal of Atmospheric and Oceanic Technology*, 26(11), 2310–2323.

Wong, D.C., Pleim, J., Mathur, R., Binkowski, F., Otte, T., Gilliam, R., Pouliot, G., Xiu, A., Young, J.O., Kang, D. (2012). WRF-CMAQ two-way coupled system with aerosol feedback: Software development and preliminary results. *Geoscientific Model Development*, 5(2), 299–312.

Wong, J., Barth, M.C., Noone, D. (2013). Evaluating a lightning parameterization based on cloud-top height for mesoscale numerical model simulations. *Geoscientific Model Development*, 6(2), 429–443.

Woodage, M.J. and Woodward, S. (2014). U.K. HiGEM: Impacts of desert dust radiative forcing in a high-resolution atmospheric GCM. *Journal of Climate*, 27(15), 5907–5928.

Woody, M.C., Baker, K.R., Hayes, P.L., Jimenez, J.L., Koo, B., Pye, H.O.T. (2016). Understanding sources of organic aerosol during CalNex-2010 using the CMAQ-VBS. *Atmospheric Chemistry and Physics*, 16(6), 4081–4100.

Wu, L., Mallet, V., Bocquet, M., Sportisse, B. (2008). A comparison study of data assimilation algorithms for ozone forecasts. *Journal of Geophysical Research: Atmospheres*, 113(D20). doi: 10.1029/2008JD009991.

Yang, B., Qian, Y., Lin, G., Leung, R., Zhang, Y. (2012). Some issues in uncertainty quantification and parameter tuning: A case study of convective parameterization scheme in the WRF regional climate model. *Atmospheric Chemistry and Physics*, 12(5), 2409–2427.

Yang, Q., Easter, R.C., Campuzano-Jost, P., Jimenez, J.L., Fast, J.D., Ghan, S.J., Wang, H., Berg, L.K., Barth, M.C., Liu, Y. et al. (2015). Aerosol transport and wet scavenging in deep convective clouds: A case study and model evaluation using a multiple passive tracer analysis approach. *Journal of Geophysical Research: Atmospheres*, 120(16), 8448–8468.

Yiou, P., Boichu, M., Vautard, R., Vrac, M., Jourdain, S., Garnier, E., Fluteau, F., Menut, L. (2014). Ensemble meteorological reconstruction using circulation analogues of 1781–1785. *Climate of the Past*, 10, 797–809.

Yu, S., Mathur, R., Pleim, J., Wong, D., Gilliam, R., Alapaty, K., Zhao, C., Liu, X. (2014). Aerosol indirect effect on the grid-scale clouds in the two-way coupled WRF-CMAQ: Model description, development, evaluation and regional analysis. *Atmospheric Chemistry and Physics*, 14(20), 11247–11285.

Zavarsky, A., Goddijn-Murphy, L., Steinhoff, T., Marandino, C.A. (2018). Bubble-mediated gas transfer and gas transfer suppression of DMS and CO2. *Journal of Geophysical Research: Atmospheres*, 123(12), 6624–6647.

Zaveri, R.A., Easter, R.C., Fast, J.D., Peters, L.K. (2008). MOdel for Simulating Aerosol Interactions and Chemistry (MOSAIC). *Journal of Geophysical Research: Atmospheres*, 113(D13). doi: 10.1029/2007JD008782.

Zhang, Y. (2008). Online-coupled meteorology and chemistry models: History, current status, and outlook. *Atmospheric Chemistry and Physics*, 8(11), 2895–2932.

Zhang, Y., Seigneur, C., Seinfeld, J.H., Jacobson, M.Z., Binkowski, F.S. (1999). Simulation of aerosol dynamics: A comparative review of algorithms used in air quality models. *Aerosol Science and Technology*, 31(6), 487–514.

Zhang, L., Gong, S., Padro, J., Barrie, L. (2001). A size-segregated particle dry deposition scheme for an atmospheric aerosol module. *Atmospheric Environment*, 35(3), 549–560.

Zhang, F., Bei, N., Nielsen-Gammon, J.W., Li, G., Zhang, R., Stuart, A., Aksoy, A. (2007). Impacts of meteorological uncertainties on ozone pollution predictability estimated through meteorological and photochemical ensemble forecasts. *Journal of Geophysical Research: Atmospheres*, 112(D4). doi: 10.1029/2006JD007429.

Zhang, H., Linford, J.C., Sandu, A., Sander, R. (2011). Chemical mechanism solvers in air quality models. *Atmosphere*, 2(3), 510–532.

Zhang, Y., Bocquet, M., Mallet, V., Seigneur, C., Baklanov, A. (2012a). Real-time air quality forecasting. Part I: History, techniques, and current status. *Atmospheric Environment*, 60, 632–655.

Zhang, Y., Bocquet, M., Mallet, V., Seigneur, C., Baklanov, A. (2012b). Real-time air quality forecasting. Part II: State of the science, current research needs, and future prospects. *Atmospheric Environment*, 60, 656–676.

Zhang, Q.J., Beekmann, M., Freney, E., Sellegri, K., Pichon, J.M., Schwarzenboeck, A., Colomb, A., Bourrianne, T., Michoud, V., Borbon, A. (2015). Formation of secondary organic aerosol in the Paris pollution plume and its impact on surrounding regions. *Atmospheric Chemistry and Physics*, 15(24), 13973–13992.

Zhang, Y., Foley, K.M., Schwede, D.B., Bash, J.O., Pinto, J.P., Dennis, R.L. (2019). A measurement-model fusion approach for improved wet deposition maps and trends. *Journal of Geophysical Research: Atmospheres*, 124(7), 4237–4251.

Zhao, C., Liu, X., Leung, L.R., Johnson, B., McFarlane, S.A., Gustafson Jr., W.I., Fast, J.D., Easter, R. (2010). The spatial distribution of mineral dust and its shortwave radiative forcing over North Africa: Modeling sensitivities to dust emissions and aerosol size treatments. *Atmospheric Chemistry and Physics*, 10(18), 8821–8838.

Zhao, A., Bollasina, M.A., Crippa, M., Stevenson, D.S. (2019). Significant climate impacts of aerosol changes driven by growth in energy use and advances in emission control technology. *Atmospheric Chemistry and Physics*, 19(23), 14517–14533.

Zheng, D., Leung, J., Lee, B., Lam, H. (2007). Data assimilation in the atmospheric dispersion model for nuclear accident assessments. *Atmospheric Environment*, 41(11), 2438–2446.

Zheng, J., Zhu, J., Wang, Z., Fang, F., Pain, C.C., Xiang, J. (2015). Towards a new multiscale air quality transport model using the fully unstructured anisotropic adaptive mesh technology of fluidity (version 4.1.9). *Geoscientific Model Development*, 8(10), 3421–3440.

Zheng, B., Tong, D., Li, M., Liu, F., Hong, C., Geng, G., Li, H., Li, X., Peng, L., Qi, J. et al. (2018). Trends in China's anthropogenic emissions since 2010 as the consequence of clean air actions. *Atmospheric Chemistry and Physics*, 18(19), 14095–14111.

Zhou, C., Zhang, X., Gong, S., Wang, Y., Xue, M. (2016). Improving aerosol interaction with clouds and precipitation in a regional chemical weather modeling system. *Atmospheric Chemistry and Physics*, 16(1), 145–160.

Zink, K., Pauling, A., Rotach, M., Vogel, H., Kaufmann, P., Clot, B. (2013). EMPOL1.0: A new parameterization of pollen emission in numerical weather prediction models. *Geoscientific Model Development*, 6, 1961–1975.

Zobler, L. (1986). A world soil file for global climate modeling. Working document, NASA Tech. Memo. 87802.

Zyryanov, D., Foret, G., Eremenko, M., Beekmann, M., Cammas, J., D'Isidoro, M., Elbern, H., Flemming, J., Friese, E., Kioutsioutkis, I. et al. (2012). 3-D evaluation of tropospheric ozone simulations by an ensemble of regional chemistry transport model. *Atmospheric Chemistry and Physics*, 12(7), 3219–3240.

Index

I, K, L

M, N

O, P, R

S, T

U, V, W

Printed and bound by CPI Group (UK) Ltd, Croydon, CR0 4YY
13/09/2026
14960997-0001